**SpringerBriefs in Applied Sciences
and Technology**

SpringerBriefs present concise summaries of cutting-edge research and practical applications across a wide spectrum of fields. Featuring compact volumes of 50 to 125 pages, the series covers a range of content from professional to academic.

Typical publications can be:

- A timely report of state-of-the art methods
- An introduction to or a manual for the application of mathematical or computer techniques
- A bridge between new research results, as published in journal articles
- A snapshot of a hot or emerging topic
- An in-depth case study
- A presentation of core concepts that students must understand in order to make independent contributions

SpringerBriefs are characterized by fast, global electronic dissemination, standard publishing contracts, standardized manuscript preparation and formatting guidelines, and expedited production schedules.

On the one hand, **SpringerBriefs in Applied Sciences and Technology** are devoted to the publication of fundamentals and applications within the different classical engineering disciplines as well as in interdisciplinary fields that recently emerged between these areas. On the other hand, as the boundary separating fundamental research and applied technology is more and more dissolving, this series is particularly open to trans-disciplinary topics between fundamental science and engineering.

Indexed by EI-Compendex, SCOPUS and Springerlink.

Maria Magdalena Ramirez-Corredores ·
Mireya R. Goldwasser ·
Eduardo Falabella de Sousa Aguiar

Decarbonization as a Route Towards Sustainable Circularity

 Springer

Maria Magdalena Ramirez-Corredores ⓘ
Chemical Sciences
Idaho National Laboratory
Idaho Falls, ID, USA

Eduardo Falabella de Sousa Aguiar ⓘ
Escola de Química
Universidade Federal do Rio de Janeiro
Rio de Janeiro, Brazil

Mireya R. Goldwasser ⓘ
Centro de Catalisis, Petroleo y
Petroquimica, Facultad de Ciencias
Universidad Central de Venezuela
Escuela de Quimica
Caracas, Venezuela

ISSN 2191-530X   ISSN 2191-5318   (electronic)
SpringerBriefs in Applied Sciences and Technology
ISBN 978-3-031-19998-1   ISBN 978-3-031-19999-8   (eBook)
https://doi.org/10.1007/978-3-031-19999-8

This Springer imprint is published by the registered company Springer Nature Switzerland AG
The registered company address is: Gewerbestrasse 11, 6330 Cham, Switzerland

Yesterday is gone. Tomorrow has not yet come. We have only today. Let us begin.
Mother Theresa

The future depends on what you do today.
Mahatma Gandhi

This book is dedicated to:
My grandkids Adrian, Miguel and Anais with
the hope that the future we had envisioned be
better for them and for society in general, by
Magdalena Ramirez,

Alberto, Leonardo, Marcos and David who
are the pillars of my life, by Mireya R.
Goldwasser, and

Christina, my wife, Felipe, my son, and
Mônica, my daughter-in-law, without whom
my life would be senseless, by Eduardo
Falabella.

Foreword

Our society is facing the urgent need for a deep change in its lifestyle, stepping away from the *consumerism* that has characterized the entire twentieth century. Moving towards a more conscious use of natural resources (that are not infinite), reducing the waste production and recycling goods is the way to go. The *Linear Economy* model (to which the *Carbon Capture and Storage-CCS* is functional) is not sustainable anymore, for both the aspects of *overexploitation of natural resources* and the *large production of waste* that have a strong impact on environment and climate. The *Circular Economy*, which implements the concepts of reducing waste by giving several lives to a good, is a more conservative approach to the use of natural resources that are finite. *Carbon Capture and Utilization-CCU* is at the heart of the Circular Economy.

Recycling of goods is an old practice, exploited at different scales for the various categories of goods. Water is used at a rate of *ca.* 4.10^{12} m^3/y and recycled at an average rate of *ca.* 1% all over the world (with some virtuous applications that recycle *ca.* 50% of the precious liquid); paper has a consumption of *ca.* 430 Mt/y and is recycled at an average rate of 50%; iron with a use of 1,000 t/y is recycled for 40%; aluminum is used at a rate of *ca.* 100 Mt/y and is recycled for 31% with some goods made of 75% recycled-Al; copper has a consumption of 20 Mt/y with a recycling rate > 45%; glass is used at rate of 205 Mt/y and recycled for 34%; plastics have a market of 360 Mt/y and a recycling of 9%; fossil-C produces *ca.* 37.10^9 t$_{CO2}$/y that has a recycling rate of only 0.6% (220 Mt$_{CO2}$/y). However, the most used goods (water and fossil-C) have the lower recycling rates.

The existing trend must be changed: we must learn from nature to recycle carbon by using atmospheric carbon dioxide, the most abundant form of carbon at hand: CO_2 is a *resource*, not a *waste*.

The decarbonization of the atmosphere is considered today as one of the key actions for mitigating the climate change. Stopping the transfer of carbon from ground to the atmosphere is the first compulsory step. But fossil-C provides 82$^+$% of the energy humanity uses today and as much of goods necessary for living. Since the end of 1700s man has progressively substituted biomass with fossil-C which has a higher energy density. Often such *working carbon* has been used with careless lavishness, believing that it will last forever, and converted into *spent carbon* released into the atmosphere. Now we know that fossil-C is a finite resource and continuing its use as we do today may cause a negative impact on the environment and climate. We have to change our attitude before we reach a non-return point. Two paths are in front of us: stopping using carbon (an economy based on hydrogen and/or electricity) or learning from Nature and develop a man-made C-cycle that may complement the natural one. Hydrogen and electricity are not the solution, they cannot substitute carbon. We can decarbonize the energy sector, we cannot decarbonize the chemical-, polymer-, pharmaceutical-, food industry. The transport sector deserves a special consideration, with some parts of it that could leave the use of C-based fuels and move to electricity and/or hydrogen (light personal transport) and other sectors that will preferentially continue to use liquid C-based fuels that are the most concentrated form of solar energy we dispose of. In particular, the aviation sector in the medium-long term will continue to use jet fuels that must move away from fossil-C towards bio-fuels or renewable fuels. A not easy task as jet-fuels must not contain oxygen and must be branched hydrocarbons (for being efficiently used at low temperatures), which rules out the currently used bio-fuels on road (bioethanol and FAMEs): a new approach is necessary that may produce in a sustainable way non-fossil jet-fuels. Most likely the choice will not be *carbon: yes or no*, but a *different way to use carbon*, in a *cyclic and sustainable mode*.

The *transition* from the *Linear to the Cyclic Economy* is not straight nor simple: it requires a lot of innovation in the production system and, thus, large investment in research.

Circularity must be sustainable. It also requires the education of people and a new social organization, a very delicate part of the play.

The Book *Decarbonization as a Route Towards Sustainable Circularity* by M. M. Ramirez-Corredores, Mireya R. Goldwasser, and Eduardo Falabella de Sousa-Aguiar makes the point on the impact of anthropogenic emissions on climate and discusses the putative remedies for alleviating the climate change, putting sustainability as a pillar of the change.

In Chap. 1, the claimed impact of CO_2 (and other Green House Gases-GHGs) emission on climate change is presented. Attention is focused on CO_2 and methane and their impact on climate change discussed in relation to the trend of their emissions. The existing parallelism between "*global population-global energy consumption*" and "*CO_2 emission-average atmospheric temperature*" is usefully shown in Fig. 1.2. Policies agreed for the reduction of atmospheric carbon are illustrated and their target for the containment of the temperature rise. CO_2 is considered responsible for climate change, as commonly reported in the literature, a role that some authors tend to share with atmospheric water vapor. It is worth to emphasize that the latter is the most abundant GHG in the atmosphere and a stronger adsorber of IR radiations than CO_2: these days it is considered by a growing number of scientists as the protagonist in temperature rise, also for the iterative effect of increase of temperature and water vaporization. Such aspect is a newcomer to the climate change problem and studies have shown that an increase of over 2%/y of the concentration of atmospheric water vapor is found in some areas of the planet. Such increase is interlinked to extreme events we observe today in areas of the planet where heavy rains were rare. The water vapor contribution to climate change is still under assessment and I believe it will be taken into due consideration in future years in making scenarios. The rise of the CO_2 atmospheric concentration is the macroscopic effect of the anthropogenic use of fossil-C and, at the end, is a way to express how much pressure humans are putting on the Earth Planet with the inefficient use of fossil carbon. In my opinion, the *cumulative effects* of the *low efficiency* in the *use of fossil-C* are responsible for the negative impact on the environment and climate, not just the released CO_2. In Chap. 1, the CO_2 molecule and its properties are also discussed together with the role of the interaction of CO_2 with metal centers and the role of catalysis in the conversion of the stable CO_2 molecule. The natural pathways for the conversion of CO_2 are examined and the dependence on the *concentration of CO_2* in the growth of plants, either C3 or C4, is analyzed.

Chapter 2 is titled *Decarbonization*, intended as the need to reduce the emission of CO_2 into the atmosphere and its accumulation by using different synergistic approaches, from increasing the efficiency in the production and use of energy to the recovery and use of CO_2. The CO_2 conversion into added value chemicals (inorganic and organic), the approaches to enhanced photosynthetic conversion, and the technical uses of CO_2 are discussed. Decarbonization is relevant to the impact of the use of fossil-C, the term has not an absolute application as biomass is still considered as a valuable source of products, materials, and fuels. Renewable carbon will be forever of great value in human activities, and CO_2 is renewable-C. The key issue is that the rate of consumption (combustion) of C-based goods is much higher (1,000– 10,000 times) than the rate of formation in natural processes. Therefore, the development of man-made processes (chemo-thermal, biotechnological, photochemical, photoelectrochemical, photo-bio-electrochemical) that have a higher rate than natural ones and can be more intensive and selective towards a target product can result an intelligent way to use CO_2. The development of a *man-made C-cycle* that complements the natural one is the key innovation for our future. Authors emphasize that governments should consider CO_2 as a resource more than a waste: this is a central point for developing the many possible uses of CO_2, from low-energy to high-energy processes. The latter require C-free primary energy sources for powering the conversion, and non-fossil hydrogen for CO_2 reduction, in a "business as usual" frame. The chapter presents various aspects of the CO_2 utilization. The production of inorganic carbonates is functional to both the long-term storage and the production of valuable goods. The barriers to the use of natural rocks as source of basic oxides that may fix CO_2 are presented through the analysis of case-study relevant to different geographic realities and economies. The possibility of using raw flue gases, carrying out a simultaneous desulphurization-decarbonation is illustrated with the relative drawbacks. The energetics and kinetics of the carbonation are analyzed and the need to develop innovative, integrated processes is emphasized. Cycling carbon in the production of sugar cane is illustrated as a case of cyclic economy. CO_2 produced in sugarcane fermentation (high purity) can be converted, by reaction with watery ammonia, into ammonium carbonate that is used to fertilize soil where sugarcane is grown. The urea synthesis, the major user of CO_2 with some $160\,Mt_{CO2}/y$ converted, is discussed with possible improvements. The valorization of CO_2 to produce organics and syngas, which can be used as energy vector or raw material for chemicals, is highlighted. Dry reforming of methane is discussed with richness of details, and the use of thermochemical cycles is examined. Co-electrolysis of CO_2 and water is the alternative to CO_2 hydrogenation for the synthesis of methanol, methane other energy products. Carboxylation of substrates is presented even considering coupling to the use of hydrogen. The fixation of CO_2 into renewable feedstock is combined with the enhanced photosynthetic conversion. The use of CO_2 as a technological fluid is summarized and the influence of CO_2-cycling on the environmental impact of heavy emitters, such as cement-, steel-, and oil-industries, is illustrated. The use of CO_2 in the medical/health sector is also shown. The chapter is closed with an analysis of the various energy scenarios and the net-zero emission option, highlighting the need of an energy-transition phase that may facilitate moving away from a fossil-C-based energy system to a decarbonized one.

Chapter 3 is centered on *Sustainability and the Circular Economy*. Carbon recycling to afford C-containing products is considered as part of the circular economy, as a way to avoid waste and is intrinsically linked to the bioeconomy. The use of renewable resources is worldwide considered as a route to mitigate the impact on the atmosphere and the climate change. Recycling-C, through CO_2 conversion, mimics Nature, avoids extraction of carbon from the ground and is a strategic part of the atmosphere decarbonization. Eco-design, eco-efficiency, and eco-effectiveness concepts are discussed. A very interesting aspect discussed is that circularity per se does not warrant sustainability: other measures need to be put in place. Closing the carbon cycle (Carbon Circular Economy-CCE) is a challenge, the most challenging aspect of circularity. The chapter also highlights the difficulty in assessing, monitoring, and measuring the implementation of the circularity strategy. Changes in business models, policies, and regulations (Governments must make a definite move towards the Recycling of Carbon and support R&I) are an essential part of the new attitude. And a key role plays the social behavior and, thus, the education of people about a new (or old?) product use. Changing the mind and behavior of people is most likely a very difficult part of the play, a net orientation of the governments in the direction of carbon recycling is the necessary step for game changing.

The entire book is a crescendo of intriguing situations that stimulate the attention of the reader and Chap. 4 is the climax. The narrative has demonstrated that technologies for large-scale conversion of CO_2 into chemicals, materials, or fuels are available at high TRL or even are under development and need research for scaling at manufacture level. Chapter 4 presents the existing international attitude to invest in CO_2 conversion as a technology for climate change mitigation through avoiding fossil-C extraction and cycling atmospheric carbon. The situation is quite diversified in the five continents, with CCS finding strong support in some areas. Obviously, CCS is not part of the Circularity strategy, being more akin to the linear economy attitude. But CCS is expected to be able to dispose of large volumes of CO_2, even if after three decades of investments in the technology worldwide only a hand of Mt/y are stored, mostly through Enhanced Oil Recovery-EOR, due to the scarce knowledge we have of potential disposal sites. The key point highlighted in this chapter is that governments must orient themselves towards carbon recycling, because this will make available large investment in R&I for developing innovation, much recommended at the various International Conventions. Some spots of light can be seen in the various multiannual programs of strong economies, such as EU, China, USA, and India. The need is to harmonize the various plans towards a common understanding that makes carbon recycling a key actor of the circular economy, through an integration with bioeconomy, converting CO_2 into a myriad of useful products using water as electron and proton donor and sunlight as a primary source of energy for powering the conversion, mimicking nature. This chapter makes a thorough analysis of possibilities and barriers to the deployment of what looks like the most obvious path to a different future. Science can develop new solutions, technology can make such solutions working, but to make them actors of the change it is necessary that policy makes the right decisions. Therefore, knowledge must flow along the chain Academia—Policymakers—Decision-makers—Industrialists—Citizens for a real game change, for real innovation permeates our society.

Chapter 4 makes the point in this important area. The dimension of the problem can be presented by three figures: $82^+\%$ of the energy is today produced from fossil-C, 37 Gt_{CO2}/y are emitted, 220 Mt_{CO2}/y are used in the chemical industry and some 30 Mt/y in technological applications. The ambition is that by 2050 fossil-C is not used anymore. The gap to fill is very large. Hydrogen cannot fill this gap alone, nor biomass. The complex problem of CO_2 mitigation requires integrated solutions. The continued use of fuels made from CO_2 and non-fossil hydrogen using perennial energy sources (solar-wind-hydro-geo) for powering the entire process may represent an intelligent support solution in the transition phase towards a defossilized energy sector.

The book *Decarbonization* demonstrates the need to change from the linear to the cyclic economy, indicates the path towards a new organization of the production system, and illustrates how our society must change its habits in order to implement an effective pro-active *modus operandi* for stabilizing or even reducing the atmospheric C content. Innovation (in any production system) plays a key role in reducing the use of fossil-C. Actually, only an average 30–32% of the chemical energy of the extracted fossil-C is used by humans, the rest is lost to the atmosphere as heat. Improving the efficiency of use of fossil-C would mean to substantially cut the CO_2 emissions for the same amount of extracted carbon. Efficiency in the production and use of any good (and energy) remains a major point for the future.

The Book *Decarbonization* is a useful text that can be easily read and interiorized even by non-specialists. I have much appreciated the way the authors have handled such a complex theme.

Bari, Italy   Michele Aresta
Innovative Catalysis for Carbon
Recycling-IC2R and Interuniversity
National Consortium on Chemical
Reactivity and Catalysis-CIRCC

Preface

The book *Decarbonization as a Route Towards Sustainable Circularity* represents a joint effort of three traditional researchers in the fields of chemical sciences and catalytic processes, aiming at identifying potential chemical routes that might be used to promote CO_2 utilization to close sustainably the carbon cycle. In this book, authors come from different countries, namely, the USA, Venezuela, and Brazil. All three have witnessed climate changes not only in their native countries, but also all over the world. Such climate changes have generated a mutual feeling that something has to be done in order to modify this scenario. Different types of actions can be undertaken to promote a better future for our beautiful blue planet; among them, education and information deserve special attention. This was the basic motivation to write our book. Our book targets a large audience that can benefit from the thorough compilation of published literature and concepts derived from not only common knowledge but also from a careful assessment of data.

The book comprises four chapters, with different goals. Chapter 1 is an introductory material that starts describing the ability of certain gases, i.e., greenhouse gases (GHGs) to absorb energy (radiative efficiency), during their effective time in the atmosphere (lifetime) causing a warming effect known as Global Warming Potential (GWP). Among the high-GWP GHGs, CO_2 is the most abundant with its concentration growing at a rate of more than 2 ppm/y. Regarding climate change, Chap. 1 also considers another concept, the Radiative Forcing (RF), which was introduced to explain the net change in the energy balance of the Earth system due to some imposed perturbation, such as the increase of GHGs atmospheric concentration. The effort of countries represented in the United Nations Framework Convention on Climate Change (UNFCCC), to achieve a temperature increase that does not exceed 1.5 °C by 2050 defined within an initiative known as the "Net-Zero" is introduced, as well as the rising of sea level and its possible effect on climate change. Emphasis is placed on the need for reduction of CO_2 emissions. This chapter also includes the description of CO_2 properties, which is a rather inert compound. The properties description and discussion are focused towards the possibilities for its storage or utilization in the manufacture of (longer lifecycle) products and/or in applications for its continuous recycle/reuse.

Chapter 2 describes pathways to convert emitted (or otherwise wasted) CO_2 into more valuable products or chemical feedstock (CU) as means for adding value or creating revenues through CO_2 utilization, which might contribute to attaining economic sustainability while solving energy and environmental issues. The need of implementing developed and emerging process technologies for the conversion of CO_2 (using low carbon energy sources) to warrant a green future is strongly stressed. Thus, inorganic as well as organic valorization routes are described, together with other more aspirational and inspirational routes, such as artificial photosynthesis, keeping in mind that CO_2 abatement cannot rely on natural photosynthesis as the unique process to reduce its concentration in the atmosphere. Process conditions, economic and energy limitations, or incentives are also incorporated in the discussion. The issues faced and addressed by the end-users and the challenges of marketing the end products are included in a separate section. Another section of the chapter introduces the net-zero initiative, approaches, and pathways to the 1.5 °C goal, to finally reach a section dealing with the potential of the different sources of energy to reach the decarbonization ambitious targets.

In Chap. 3, different aspects of a sustainable circular economy are discussed, by reviewing the cumulative knowledge on strategies, approaches, and business models developed to close the cycle involving C-bearing compounds and materials. Some approaches, such as the role of the bioeconomy and the renewable resources, as well as the introduction of the eco-design, eco-efficiency, and eco-effectiveness concepts, are discussed. A Circular Economy (CE) conceived as a way of minimizing, even avoiding waste generation is most appropriate in a decarbonizing scenario. However, circularity per se is not a warrant for sustainability and other measures need to be in place, to accomplish a sustainable circular economy. Within a decarbonization strategy, closing the carbon cycle becomes a circularity challenge. Unless the electric sector firstly passes through a drastic decarbonization, any electrification strategy will fail or be too slow to achieve the sustainability goals. The difficulties in assessing, monitoring, and measuring advances of circularity are also part of this chapter. Other required changes, such as business models, policies and regulations, social behavior, and product uses, for instance, are also discussed.

Finally, a discussion of the importance of a change in social behavior that needs to be guided by the principles developed through the scientific investigation and findings to analyze the environmental threat is presented in Chap. 4. The difference in magnitude between emissions (Gton) and utilization (Mton) is a measure of the required efforts for developing solutions. In previous chapters, the basis to claim that there is no universal/single solution to the emissions problem was set. Multiple technologies are required and need to be developed, within a holistic set of criteria. The needs, challenges, and existing gaps identified in the CU advances discussed in Chap. 2 are collected in this chapter. Implementation of technologies should take place, by circular integration, within circular business models. A vision of the aspirational (better) future is also offered in this chapter.

The authors have the strong conviction that a brighter future for our planet may be achieved provided convenient actions are undertaken. Such actions depend not only on governments' policies but also on the conscientization of the population. Conscientization is a word that conveys the idea of developing, strengthening, and changing consciousness. Conscientization requires education; however, it may change an existing culture. After all, we have a firm belief on what the great Chilean poet Pablo Neruda once said:

> *We'll slowly*
> *solve everything:*
> *we'll force you, sea,*
> *we'll force you, earth*
> *perform miracles,*
> *because in our very selves,*
> *in the struggle,*
> *is fish, is bread,*
> *is the miracle.*

Rio de Janeiro, Brazil   Eduardo Falabella de Sousa Aguiar
Idaho Falls, USA   Maria Magdalena Ramirez-Corredores
Caracas, Venezuela   Mireya R. Goldwasser

Acknowledgement

One of the authors, Ramirez-Corredores acknowledges the support for the elaboration of this manuscript to the Directed Research & Development (LDRD) Program of Battelle Energy Alliance, LLC under DOE Idaho Operations Office contract No. DE-AC07-05ID14517. The United States Government retains and the publisher, by accepting the manuscript for publication, acknowledges that the United States Government retains a nonexclusive, paid-up, irrevocable, worldwide license to publish or reproduce the published form of this manuscript or allow others to do so, for United States Government purposes.

Highlights and Graphic Abstract

Highlights

- Global carbon dioxide emissions are considered the primary contributor to climate change.
- Carbon, one of the most abundant elements on the planet can sustain a thriving economy if used and managed with scarcity criteria.
- Implementation of decarbonizing strategies is an urgent need for main stakeholders.
- The connection between a decarbonizing strategy and a sustainable circularity should involve carbon dioxide valorization.
- Sustainable circularity passes through a transition that converts any and every supply chain into a value chain.
- CO_2 utilization is an enabler for a sustainable circular economy.

Graphic Abstract

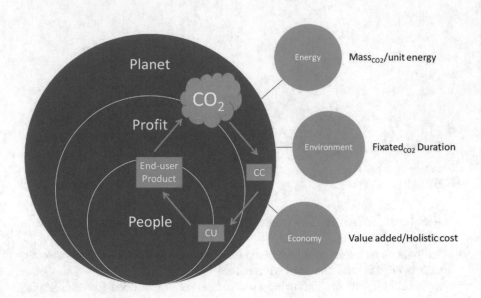

Contents

Abbreviations and Acronyms

A/R	Afforestation and reforestation
BC	Biochar
BECCS	Bioenergy with carbon capture and storage
CBE	Circular bioeconomy
CBM	Circular Business Models
CC	Carbon capture
CCC	Cambridge carbon capture
CCP	Compensation point
CCS	Carbon capture and storage
CCSM	Carbon capture and storage by mineralization
CCU	Carbon capture and utilization
CCUS	Carbon capture, utilization, and storage
CDM	Clean development mechanism
CDR	Carbon dioxide removal
CE	Circular economy
CETS	Chemical energy transmission systems
CLC	Chemical-looping combustion
CLEA	Closed loop efficiency analysis
COP2X	2Xth. Conference of the (United Nations) parties
CS	Carbon storage
CSP	Concentrated solar power
CU	Carbon utilization
DAC	Direct air capture
DACCS	Direct air carbon capture and storage
DMC	Dimethyl carbonate
DME	Dimethyl ether
EnPIs	Energy performance indicators
EOR	Enhanced Oil Recovery
ERF	Effective Radiative Forcing
EROEI	Energy returned on the energy invested
EU	European Union

EV	Electric vehicles
EW	Enhanced weathering on land and in oceans
FE	Faradaic efficiency
FEEM	Fondazione Eni Enrico Mattei
GC	Glycerol carbonate
GCS	Geological carbon storage
GDE	Gas diffusion electrodes
GDP	Gross domestic product
gge	Gallon of gasoline equivalent
GHG	Greenhouse gas
GMSL	Global mean sea level
GMST	Global mean surface temperature
GWP	Global Warming Potential
HFO	Heavy fuel oil
HTE	High temperature electrolysis
IAM	Integrated assessment models
IAW	Industrial alkaline waste
IEA	International Energy Agency
IES	Integrated energy systems
IGCC	Integrated gasification combined cycle
IMO	International Maritime Organization
IPCC	Intergovernmental Panel on Climate Change
LCA	Life cycle assessment
LCFS	Low carbon fuels standard
LHV	Lower heating value
LNG	Liquefied natural gas
LPG	Liquified petroleum gas
LTE	Low temperature electrolysis
MC	Mineral carbonation
MCCU	Mineral carbon capture and utilization
MCPs	Materials, components and products
MDO	Marine diesel oil
MDR	Methane dry reforming
Me	Metal
MPO	Methane partial oxidation
MR	Membrane reactor
MSW	Municipal solid wastes
MTO	Methanol to olefins
NET	Negative emissions technology
NETL	National Energy Technology Laboratory
NG	Natural gas
NOAA	National Oceanic and Atmospheric Administration
NPP	Nuclear power plants
NTP	Non-thermal plasma
OCF	Oxidized carbon-felt disks

OCM	Oxidative coupling of methane
OECD	Organization for Economic Co-operation and Development
OF	Ocean fertilization
OPC	Ordinary Portland cement
PCC	Post-combustion carbon capture
PLA	Polylactic acid
PtG	Power-to-Gas
PV	Solar photovoltaic
RD&D	Research, development and deployment
RE	Renewable energy
RF	Radiative Forcing
RJF	Renewable jet fuel
RRfW	Resource recovery from waste
rWGS	Reverse water gas shift
SAF	Sustainable aviation fuel
SC-CO_2	Supercritical carbon dioxide
SCE	Sustainable circular economy
SCS	Soil carbon sequestration
SDG	Sustainable Development Goal
SDS	Sustainable development scenario
SDSN	Sustainable development solutions network
SMR	Methane steam reforming
SNG	Synthetic or substitute natural gas
SOEC	Solid oxide electrochemical cell
SPARG	Sulfur passivated reforming
STEPS	Stated policy scenario
TEA	Techno-economic analysis
TIMES	The integrated Markal-Efom system
TPE-CMP	Thermoplastic elastomer-Conjugated microporous polymer
TREMP	Topsoe recycle energy-efficient methanation process
UK	United Kingdom
UN	United Nations
UNEP	United Nations Environment Program
UNFCCC	United Nations Framework Convention on Climate Change
US	United States
WGS	Water gas shift
WtU	Waste-to-Urea

List of Figures

Chapter 1
Carbon Dioxide and Climate Change

1.1 Greenhouse Gases and Climate Change

Historically, the progress of humanity has advanced based on the mechanization/automation of processes that significantly lead to speed changes in developments that satisfy its needs. In this context, industrial progress led to the use of fossil fuels (coal, oil, and gas) which has contributed positively to its development. Even though the use of fossil fuels as an energy source to improve people's welfare is undeniable, its production and consumption have been thoroughly linked to the degradation of the environment. The emission of contaminating gases into the atmosphere has been steadily changing climatic conditions. The global climate change is closely connected to the emission of anthropogenic greenhouse gases (GHGs). GHGs can be seen as an insulating layer that refrains the Earth's energy from escaping to space and absorbs energy from space and transfers it to the Earth's surface, causing a net warming effect. This warming effect known as Global Warming Potential (GWP) differs among the various GHGs, depending on their ability to absorb energy (radiative efficiency), and on their effective time in the atmosphere (lifetime). The GWP definition allows a normalized comparison among the GHGs since it measures how much energy the emissions of 1 ton of a given gas will absorb over a given period of time, relative to the emissions of 1 ton of CO_2. The usual period of time is 100 years and the GWP of CO_2 is established as 1, regardless of the time period employed. For instance, the CH_4 radiative efficiency is greater than that of CO_2, its lifetime is much shorter (12 ± 3 years), making its GWP 28–36/100 y [1]. Nevertheless, GWPs are periodically updated with better estimates of the radiative efficiency, lifetime, and atmospheric concentration.

The GWP of GHGs has been held responsible for the observed continuous and steady rise in global temperature. In fact, as reported by the Intergovernmental Panel on Climate Change (IPCC) [2], the observed mean land surface air temperature has increased noticeably more than the global mean surface (land and ocean) temperature

© The Author(s), under exclusive license to Springer Nature Switzerland AG 2023
M. M. Ramirez-Corredores et al., *Decarbonization as a Route Towards Sustainable Circularity*, SpringerBriefs in Applied Sciences and Technology,
https://doi.org/10.1007/978-3-031-19999-8_1

(GMST[1]), since the pre-industrial period (1850–1900).[2] The warming induced by human activities reached nearly 1 °C above pre-industrial levels in 2017, and it is expected to keep increasing at a rate of between 0.1 and 0.3 °C per decade reaching 1.5 °C, by 2040 [3]. A little over 3 °C, by 2100 has been projected under this "business as usual" scenario [4]. Therefore, stronger and deeper mitigation and abatement measures, actions and policies need to be in place and start execution immediately.

Another concept to consider regarding climate change is Radiative Forcing (RF), which was introduced to explain the net change in the energy balance of the Earth system due to some imposed perturbation, such as the increase of GHGs atmospheric concentration. The RF refers to a change in net (down minus up) radiative flux (shortwave plus longwave; in W/m^2) due to an imposed change, over a particular period of time and quantifies the corresponding energy imbalance. In order to counteract these flux changes, the Earth's systems respond through climate changes, and all such responses are explicitly excluded from this definition of forcing. The climate sensitivity parameter, λ establishes an assumed relation between a sustained RF and the equilibrium GMST response (ΔGMST), as per $\lambda = \Delta\text{GMST/RF}$. Similarly, an Effective Radiative Forcing (ERF) is defined as the change in net (top of the atmosphere) downward radiative flux after allowing for atmospheric temperatures, water vapor and clouds to adjust, but with surface temperature or a portion of surface conditions unchanged. Although water vapor is the primary GHG in the atmosphere and its warming effect is approximately two to three times greater than that of CO_2, its condensation and subsequent precipitation as rain, snow or hail minimizes its residence time. Hence, water vapor has a negligible impact on overall atmospheric concentrations and RF and does not contribute significantly to the long-term greenhouse effect.

The most abundant GHGs populating the Earth atmosphere are carbon dioxide (CO_2), methane (CH_4), water vapor (H_2O), nitrogen oxides (NO_x), chlorofluorocarbons (CFCs), perfluorocarbons (PFCs), sulfur hexafluoride (SF_6), and ozone (O_3). Some of these gases emerge from natural processes, but the problems start when emissions are provoked by anthropogenic sources and activities. In fact, human activities have led to 40% increase in the CO_2 concentration in the atmosphere from 280 ppm in 1750 (initiation of the industrial revolution) to 417 ppm by the end of 2021 (also the same on March 17, 2022). The current uncertainty in the National Oceanic and Atmospheric Administration (NOAA) determination of the CO_2 concentration is about 0.18 ppm [5]. The yearly growth of the atmospheric concentration from 2020 to 2021 was 2.14 ppm [6], as has been the trend through the last 7 decades (see Fig. 1.1).

---

[1] GMST is the estimated global average of near-surface air temperatures over land and sea ice, and sea surface temperatures over ice-free ocean regions, with changes normally expressed as departures from a value over a specified reference period. When estimating changes in GMST, near-surface air temperature over both land and oceans are also used.

[2] The pre-industrial period of 1850–1900 was set by the Intergovernmental Panel on Climate Change (IPCC) to approximate pre-industrial GMST that would have been observed during the multi-century period prior to the onset of large-scale industrial activity around 1750.

Fig. 1.1 CO$_2$ emissions growth since second half XX century (data from Ref. [6])

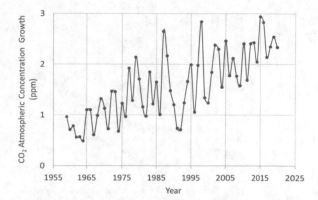

Thus, CO$_2$ represents the largest part of the emitted GHGs, and it has been the case since the modern era, converting CO$_2$ into a silent enemy of humanity with a high impact in pollution. The problem began to alarm when CO$_2$, the common residue of fossil fuels, was made directly responsible for global warming, threatening human health and the quality of life at short-term, and affecting ecological balance and biological diversity in the long-term. Since individual organisms are adapted to live within specific ranges of climate conditions (e.g., temperature, precipitation, humidity, sunlight, etc.), changes in these climate conditions will affect the health and function of ecosystems and might threaten the survival of entire species. All forms of life (microbes, plants, animals, and humans) are an intrinsic and integral part of the carbon cycle that mediates and modifies the Earth's climate. The examination of geological records shows how life has been driving the chemical composition of the atmosphere through time. However, the question on whether CO$_2$ is directly responsible or is just an indicator of the impact caused by human activities on the atmosphere has been posted. The fact of the inefficient conversion of chemical energy into other forms of energy and their inefficient uses and applications justifies such a question. Currently, chemical energy is converted either to heat, mechanical or electricity with efficiency in the range of 27–35%, releasing typically 73–65% of the original chemical energy to the atmosphere in the form of heat. Therefore, is it the growing CO$_2$ concentration in the atmosphere, or the dissipated heat responsible for global warming? Or both? [7]. The 1/3 ratio between these two factors rises a concern on where attention should be focused, i.e., emissions vs. efficiency. The close relationship between the two points towards consolidated efforts on both and addressing all efficiency related issues around electricity generation, heating systems, industrial plants, and the transport sector since these are intrinsically connected with human activities and life. The connection between energy efficiency (in both cases, its generation and its application, consumption or uses) and CO$_2$ emissions and their impact in climate change rise further more concerns when one observes the rate of population growth [8] and the consequent increase in energy consumption [9], as shown in Fig. 1.2.

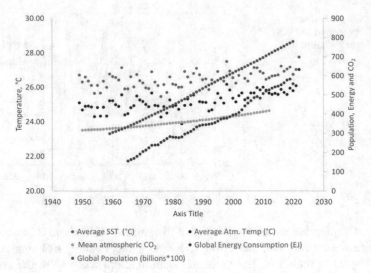

Fig. 1.2 Interconnected climatic factors

The observed atmospheric global average temperature and the global average surface sea temperature [10], as well as the atmospheric concentration of CO_2 [6], were also included in Fig. 1.2. This plot indicates that the increase rate of population growth parallels well that of energy consumption, while the rate of increase in the global temperatures parallels that of the atmospheric CO_2 concentration.

Climate systems and energy security (needed for human development) are intrinsically connected to carbon emitting power plants particularly and to renewable energy (RE) systems [11]. In the past two centuries, the atmospheric concentration of CO_2 has increased considerably, with global emissions expanding in the 30–35 Gton range [12, 13], from which more than 5 Gton corresponded to the US [14]. The emissions decrease observed in 2020 was explained as due to the economic recession caused by the COVID-19 pandemic. Nevertheless, a continuous growth of 0.43 Gton/y up to 2040 is expected from now on, as economy gains track again.

Coordinated from the parties involved in the UNFCC, different strategies, action plans, and economic tools have been suggested and implemented around the world. Chronologically, in 1997 the first international treaty known as the Kyoto Protocol was subscribed with the objective of attaining by 2012, 5% GHGs emissions reduction, from the levels observed by 1990. In order to account for economic differences among the subscribing parties, the Protocol implemented three flexibility mechanisms, namely the international emissions trading, the clean development mechanism, and the joint implementation. Then, in 2010, the UNFCCC offered the Cancun Agreement to 76 countries to voluntarily pledge to control GHG emissions. The involved countries were responsible for 85% of annual CO_2 emissions at that time. In 2011, the implementation of carbon capture and storage (CCS) technologies was recommended by the Durban Agreement as eligible projects and activities within the clean development mechanism (CDM). Unfortunately, by 2011 the demonstrated CO_2 geological storage was facing concerns on risks, leakage, environmental impacts, and public acceptance. The most remarkable agreement was reached in Paris, by 2015 when 196 countries subscribed it. Nevertheless, the Paris Agreement could only become legally binding if 55% of the subscribing countries ratified it. Although the Paris Agreement called for zero net anthropogenic GHG emissions, the European Union suggested at least a 40% reduction of the 1990 baseline emissions, by 2030. Clearly, one of the goals of the Paris Agreement was to reduce CO_2 emissions to counterbalance its growing tendency [15]. The Paris Agreement also introduced the term *"carbon budget"* to define the amount of CO_2 that humanity can emit while still having a chance to contain global warming. Thus, carbon budget monitoring and control is the center of attention when trying to achieve climate change mitigation goals. In 2021, with the signing of the Glasgow Climate Pact by all the states belonging to the United Nations (UN) during the 26th UN Climate Conference [16], a commitment was achieved for the first time to eliminate emissions of GHGs of anthropogenic origin especially CO_2. This Pact completed the Paris Agreement's rulebook, by establishing market mechanisms and non-market approaches, as well as a transparent reporting of the climate actions and support provided or received, including for loss and damage. Thus, at present, *all UN countries have agreed to work in reducing the gap between existing emission reduction plans and what is required to reduce emissions, so that the rise in the global average temperature can be limited to 1.5 °C, compared with pre-industrial levels*.

Although there is a certain natural capacity for sinking carbon emissions in forests and oceans, the rate at which GHGs are emitted overpasses such capacity, leading to an atmospheric accumulation that threatens environmental sustainability. In fact, in the data revealed by the Global Carbon Project [17], CO_2 absorption by terrestrial ecosystems was 12.5 Gton/y, and 9.2 Gton/y by oceans, during the 2010's decade, which represented an average of around 60% compensation for the emissions. Meanwhile, in 2020, 54% were compensated by these two massive sinks. This project has estimated that the remaining carbon budget for the next 20 years and assuming emissions for 2022 at the same level to those of 2021, for a 50% likelihood of limiting global warming to 1.5 °C was 770 $Gton_{CO2}$.

Finally, decades of observations and research, based on basic principles of physics and chemistry led scientists to hypothesize that burning fossil fuels will contribute to increase in the Earth's average surface temperature. This understanding was extended to identify other affecting factors, such as land-use changes and to establish modes of action and interactions that modulate the long-term warming trend. Processes, scenarios, and models were studied and considered for projecting climate changes into the future. Multi-disciplinary teams have explained the interactions of natural as well as human systems (e.g., agricultural systems, ecosystems, water resources, etc.) with the climate system. Thus, throughout the years, a strong, credible body of evidence, based on multiple lines of research has been collected, documenting changes in the climate and supporting that these changes are at large partly caused by human activities. Additionally, climate change poses a substantial risk for many human and natural systems. A factual summary [18] includes:

(i)   The Earth's average surface temperature was 0.8 °C warmer during the first
       decade of the twenty-first century than it was during the first decade of the
       twentieth century,
(ii)   Most of the warming has been attributed to human activities that release heat-
       trapping GHGs into the atmosphere, especially CO_2,
(iii) Natural climate variability can explain short-term climate changes, as well as
       regional differences; however, it cannot account for the long-term warming
       trend,
(iv)   Climate changes also include increases in frequency of intense rainfall,
       decreases in Northern Hemisphere snow cover and Arctic-sea ice, more
       frequent warmer days and nights, rising sea levels, and widespread ocean
       acidification,
(v)   Human and natural systems at risk include freshwater resources, the littoral
       environment, ecosystems, agriculture, fisheries, human health, and national
       security, among others,
(vi)   The magnitude of climate change and the severity of its impacts depend strongly
       on the measures taken and implemented.

Regardless of the magnitude of the volume of scientific knowledge and collected
evidence supporting climate change, science is not exempt from uncertainty. At
this point, it is worth to clarify that scientific uncertainty describes the precision
and/or confidence of an experimental measurement, an observation and even a created
knowledge [19]. The social interest and implications on society of climate change
have pushed the scientific community to develop explicit techniques, methods and
methodologies for the assessment of the uncertainty on any new theory or conclusion
reached in the area [20].

The rising of sea levels is probably one of the best measured indications of
the impact of climate change. Sea level has been systematically assessed by tide
gauges for more than 100 years and it has risen more than 120 m since the Ice Age
(26,000 years ago), with a relatively steady rise over the past 6,000 years [21, 22]. The
average rate of sea level rise was 1.3 mm/y between 1901 and 1971, then it increased
to 1.9 mm/y between 1971 and 2006, with a further increase of 3.7 mm/y between
2006 and 2018 [23]. Any and every modeled scenario predicts further increases
during the 21st Century, the likely global mean sea level (GMSL) rise by 2100
relative to 1995–2014 would be in the range from 0.28 m for a very low GHG
emissions scenario up to 1.01 m under the very high GHG emissions scenario [18].
Two fundamental processes cause the rise of the sea level: (i) the thermal expan-
sion of the water mass of the ocean basins as it absorbs heat (ocean temperature
and heat content), and (ii) the addition of water from land-based sources, due to
the melting of ice sheets and glaciers (water mass). These two processes are highly
accelerated by global warming, turning into a higher negative impact on sea level
[24]. However, other influencing factors include ocean salinity-acidity, water mass
properties, surface fluxes, and winds/ocean circulation.

1.2 Carbon Dioxide

Climate change and air quality are major environmental concerns because they directly affect life on Earth. CO_2 is the side-product, waste stream or residue of several human activities such as burning of fossil fuels, cement manufacture, fermentation, industrial processes, among others, causing a greenhouse effect by trapping some of the sun's heat, leading to an increase of Earth's temperature, contributing to climate change and issues such as melting glaciers, increase in droughts, storms, heat waves, floods, and cyclones [21]. This situation has prompted the international community to dictate a series of regulations that affect and decrease the activities of the petroleum industry, introducing legal statutes that have to be accomplished by conventional fuels in the short-term, in order to preserve the environment [25]. It is expected that the pressure to significantly reduce the use of world oil reserves will also trigger some international conflicts over access to natural resources, mainly of fossil origin, causing serious economic crisis worldwide, especially in countries without technologies developed towards the generation of alternative energy, thereby exposing the world to geopolitical risk. It is important to highlight the fact that, right now, in the 2021 winter season, several European countries are having serious economic problems due to the high increase in electricity prices, especially the gas sector. This situation becomes more and more evident, when the emergence of both multinational and multilateral initiatives is observed, calling to diversify the current energy matrix through the development and use of renewable alternative energies, friendly to the environment [13, 26] and/or technologies that mitigate the environmental impact of the current matrix [27–30].

CO_2 is a relatively simple nontoxic linear molecule, with a formula of a carbon atom linked to two oxygen atoms. It occurs naturally in Earth's atmosphere as a trace gas. The current CO_2 concentration in the atmosphere is about 0.04% (417 ppm), reaching levels 50% higher than before the industrial revolution (280 ppm) [31]. CO_2 is one of the most abundant renewable carbon resources in nature, it occurs from volcanoes, forest fires, hot springs, and geysers, being also freed from carbonate rocks by dissolution in water and acids [32]. Other sources include outgassing from the ocean, decomposing vegetation and other biomass, and even flatulence from ruminant animals. Because carbon dioxide is soluble in water, it occurs naturally in groundwater, rivers and lakes, ice caps, glaciers, and seawater. Also, it is present in occurrences of petroleum and natural gas (NG) [33]. It is a neutral, colorless, slightly acidic noninflammable, which under normal conditions of temperature and pressure, is in a gaseous form with a high thermodynamic stability ($\Delta G^{\circ}{}_f = -396$ kJ/mol). At temperatures below $-79\,^{\circ}C$, it can solidify and liquefy when dissolved in water. The thermodynamic stability of CO_2 results in the use of high temperatures for conversion reactions, leading to processes with high energy intensity.

CO_2 decomposes at temperatures above 2000 °C, producing toxic fumes of carbon monoxide and reacting violently with strong bases and alkali metals. The low reactivity of CO_2 makes it to be perceived as rather inert and gives rise to physical applications as well as for chemical uses as solvent or as inert reacting media. However,

Table 1.1 Selected set of physical, chemical, thermodynamic, and physicochemical properties of CO_2

Property	Value
Molar mass (g/mol)	44.01
Boiling point: (°C)	−78.46
Density (kg/m^3)	1.778
Specific entropy (kJ/kg·K)	2.753
Std molar entropy (S°$_{298}$, J/mol·K)	214
Specific enthalpy (kJ/kg)	510.09
Std enthalpy of formation ($\Delta H_f°_{298}$, kJ/mol)	−393.5
Heat (enthalpy) of fusion (kJ/mol)	9.02
Heat (enthalpy) of sublimation (kJ/mol, @ 180 K)	26
Heat (enthalpy) of evaporation (kJ/mol @ 15 °C)	16.7
Heat capacity (J/K·mol)	37.135
Thermal conductivity (W/m·K @ 300 K)	0.017
Dynamic viscosity (cP)	1.495
Kinematic viscosity (cSt)	0.834
Thermal diffusivity (m^2/s)	$1.132 * 10^{-5}$
Solubility in water g/L at 25 °C and 100 kPa	1.45
Acidity (pKa, 1 and 2)	6.35, 10.33
Ionization potential (eV)	13.77

the CO_2 chemical conversion limitations associated with this assumed inertia may be underestimated when realizing it is isoelectronic to highly reactive molecules such as isocyanates and ketenes. In Table 1.1, a set of selected properties of CO_2 are collected, particularly those, which constitute the basis for concepts of its uses, applications, and conversion.

The critical temperature of CO_2 is 31.06 °C and the pressure is 73.8 bars (1070 lb/in^2), at which it exhibits a critical density of 0.469 g/cm^3. Under these conditions, CO_2 shows excellent solvency power and several extraction processes have been developed for its use as an extractant, under supercritical conditions. Compared to air, the density of CO_2, at standard temperature and pressure is 1.67 times greater (1.98 kg/m^3), making it difficult to diffuse into the atmosphere and causing the observed accumulation, towards Earth surface.

The incorporation of CO_2 as feedstock, in chemical processes, is considered both environmentally and economically attractive and a valuable contribution to managing carbon species, which can further be the basis of a sustainable carbon-economy. The above defined properties of CO_2 indicate broad possibilities for numerous applications, additionally to the currently known as a solvent for extraction processes, such as caffeine extraction [34]. Most of the CO_2 reactions are endothermic and only a few are exothermic. Even for these few exothermic reactions, most of the overall processes are endergonic, typically due to the entropy change between products and reagents. Thus, the first step is CO_2 activation, by the addition of energy followed by the reaction of the activated CO_2 intermediate to form further energy-rich intermediate streams or products. These intermediate streams could be used for transferring the CO_2 moiety to its final fixation in longer lifecycle products. Clearly, process intensification through advanced reaction engineering solutions, such as the integration of the reaction and separation processes, might be a valid approach for overcoming this issue. The conversion reactions of CO_2 proceed through two different pathways: a reductive pathway and an acid–base pathway. The reduction reaction is energy intense, and catalysts are typically employed to diminish the energy requirements. Whether chemical or biological catalysts are used, additional means have also been studied, such as electrochemistry and photochemistry. Thus, the reductive transformation of CO_2 into value-added products can be achieved by single or multiple combinations of thermal, chemical, catalytic, biological, electrolysis, and photo-induced means (e.g., photoelectrobiothermocatalytic).

The low reactivity of CO_2 places catalysis in a very important role since more efficient and effective catalysts are needed to promote and control the desired reactions, enabling the development of successful process technologies. Furthermore, better catalysts can only be attained through a thorough knowledge of the reaction pathways and mechanisms, together with the identification of barriers and technical limitations [35]. The role of catalysis in developing opportunities for the combined CO_2-H_2 pair has been discussed in Ref. [36]. Nevertheless, as mentioned in the previous paragraph, other sources of activation might contribute to enhance catalytic efficiency.

An explanation of the electrophilic and nucleophilic reactivity of CO_2 is represented in Fig. 1.3 [37]. This figure depicts a threefold reactivity, which is provided by the nucleophilic oxygen atoms, the electrophilic carbon atom and the π system, and the coordination chemistry to metal centers.

The acid–base reactivity of CO_2 has been advantageously used in capturing, storing and sequestration. Quite often, such routes take advantage of the fact that although CO_2 is very stable in oxidation reactions, CO_2 is an anhydride, or rather,

Fig. 1.3 CO_2 reactivity and examples of catalytic activation (Reproduced from Ref. [37], under unrestricted Beilstein-Institut Open Access License Agreement 1.2)

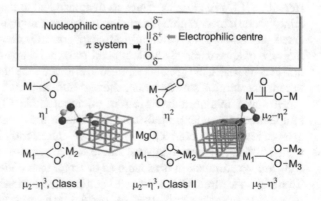

Table 1.2 Examples of reactions involved in the natural CO_2 cycle

		Time scale (y)
1.1	$CO_2 + CO_3{}^{2-} + H_2O \rightarrow 2HCO_3{}^{-}$	$10-10^3$
1.2	$6CO_2 + 6H_2O + photons \rightarrow C_6H_{12}O_6 + 6O_2$	$1-10^2$
	$C_6H_{12}O_6 + 6O_2 \rightarrow 6CO_2 + 6H_2O + heat$	
1.3	$CO_2 + CaSiO_3 \rightarrow CaCO_3 + SiO_2$	10^4-10^6
1.4	$CO_2 + CaCO_3 + H_2O \rightarrow Ca^{2+} + 2HCO_3{}^{-}$	10^3-10^4

it is an acidic oxide. Thus, a solution of CO_2 in water is acidic and the higher the CO_2 concentration, the higher the acidity would be. This property confers characteristics to this molecule, allowing it to undergo several types of reactions, particularly with alkaline and alkaline earth elements as well as basic oxides. The formation of carbonates, bicarbonates and carbamates routes as CO_2 valorization alternatives will be discussed in detail in Chap. 2.

The natural CO_2 cycle leads to a molecular exchange between the atmosphere and Earth's surface (land and ocean). This exchange is kinetically controlled (see the time scale of their occurrence in Table 1.2), by the undergoing chemical reactions, e.g., solubility in sea water (buffering, reaction 1.1), photosynthesis (and respiration, reaction 1.2), weathering (mineralization, reaction 1.3), and carbonate dissolution (reaction 1.4).

Thus, rapid exchange through the biosphere or living beings expands from years to centuries and becomes slower when moving into the oceans (decades to multicentennial) or deeper to aquifer systems and subsurface layers (millions of years). The buffering capacity of the oceans decreases at higher CO_2 concentrations. Consequently, the larger the cumulative atmospheric concentration, the lower the proportion uptaken by the oceans and the higher the remaining atmospheric fraction. Reactions 1.2 summarize the set of reactions known as Carbon Fixation. Carbon fixation is the process wherein photosynthetic organisms (such as plants) turn inorganic carbon into organic compounds (carbohydrates). CO_2 fixation, for instance, is a type of carbon fixation wherein carbon dioxide from the atmosphere is converted into carbohydrates. Regarding CO_2 fixation in plants, as demonstrated in reactions 1.2, the role of CO_2 atmospheric concentration is a matter of dispute. Previous studies [38, 39] have indicated that a 300 ppm increase in atmospheric CO_2 concentration brings about an increase of approximately 30% in plant growth. Even when plants are cultivated in nutrient-poor soil, the growth response to elevated CO_2 has been large, in comparison to nutrient solution studies which showed little response.

However, this stimulating effect of atmospheric CO_2 enrichment is strongly temperature dependent. Studies suggest that if the climate warms up, the average growth response to doubled CO_2 could be consistently higher than the aforementioned 30%. Furthermore, several studies have suggested that under water-stress, the CO_2 growth stimulation is as large as or larger than under well-watered conditions. Therefore, the direct CO_2 effect will compensate somewhat, if not completely, for a hotter, drier climate. For that reason, predicting the ultimate biosphere consequences of a doubling of the Earth's atmospheric CO_2 concentration seems to be much more complex than what has been originally anticipated. Data reported and depicted in a plot by Rosenthal [40] suggest that the photosynthesis process may virtually stop, at a CO_2 concentration of around 200 ppm.

The influence of atmospheric CO_2 concentration is related to the concept of CO_2 compensation point (CCP). The CCP is the CO_2 concentration at which the rate of photosynthesis matches exactly the rate of respiration. It depends on the type of plants (C3 or C4). C3 plants are defined as the plants that exhibit the C3 pathway. These plants use the Calvin cycle in the dark reaction of photosynthesis. On the other hand, C4 plants are defined as the plants that use the C4 pathway or Hatch-Slack pathway during the dark reaction. The Calvin cycle is part of a photosynthesis that occurs in two stages. In the first stage, chemical reactions use energy from light to produce ATP and NADPH. In the second stage (Carbon cycle or dark reactions), CO_2 and H_2O are converted into organic molecules, such as glucose. In Hatch-Slack views, the C4 pathway involves steps that first convert pyruvate to phosphoenolpyruvate (PEP) to bind with the CO_2 forming a four-carbon compound (hence the name C4). As a result, the photorespiration pathway is bypassed, and the wasteful loss of CO_2 common in C3 carbon fixation pathway is minimized. About 95% of the plant species on the planet are C3, including rice, wheat, soybeans, and all trees.

There is a significant difference in CCP between C3 plants and C4 plants: on land, the typical value for CCP in a C3 plant ranges from 40–100 μmol/mol, while in C4 plants the values are lower at 3–10 μmol/mol. Hence, increasing C_{O2} levels in the atmosphere enhances the production of C3 plants [41].

Moreover, it was observed that the photosynthesis of wheat and maize decreased when such crops were fertilized with NH_4^+ as a nitrogen (N) source at ambient CO_2 concentration in comparison to those grown with a mixture of NO_3^- and NH_4^+, or NO_3^- as the unique N source. Customized chambers were used to grow the test plants aiming at maintaining a continuous and desired atmospheric CO_2 (Ca) concentration: 280 μmol/mol (representing the pre-Industrial Revolution CO_2 concentration of the eighteenth century), 400 μmol/mol (present level) and 550 μmol/mol (representing the potential future concentration in 2050). Results indicate that future increases in atmospheric CO_2 make C3 plants more likely to use NH_4^+ rather than counting on NO_3^- as a source of N fertilizers for crop production.

A rather disseminated theory associated with atmosphere–biosphere interactions is the Gaia hypothesis of Lovelock [42]. Such hypothesis presumes that the rate of CO_2 removal from the atmosphere by C3 plants decreases with the decreasing CO_2 concentration, thereby acting as a self-regulatory check on atmospheric CO_2 depletion. An important premise of the Gaia hypothesis is that the CCP of C3 primary productivity is near 150 mmol mol^{-1} [43].

The lowest CO_2 concentration in the atmosphere to allow plant growth is indeed a matter of dispute and a rather controversial topic. Undoubtedly, very low CO_2 concentrations severely retard growth, but do not stop it completely. Additionally, CO_2 concentration also affects the mortality rate. Seedling establishment is a critical stage in plant growth whose main mortality agents are environmental stress, especially drought, disease and predation, and heat stress. Low CO_2 concentrations would greatly increase the time during which a casual stress event could negatively influence the seedling. When stress takes place, its severity is greater at low CO_2 concentrations, since such low concentrations are known to intensify both abiotic and biotic stress. On the other hand, high CO_2 concentrations alleviate the effects of many forms of stress. Notwithstanding, there are some evidence that these slowly growing seedlings and the mortality thereof are more vulnerable to stochastic events [44].

The dynamic changes in CO_2 concentration in the atmosphere, land and oceans give rise to temporal imbalances. As it was mentioned above, the growth of atmospheric CO_2 concentration has been increasing with time, while the CO_2 absorption capacity of forests and oceans is decreasing with time. In the long-term, these imbalances lead to shifts in the carbon cycle since the relative rate of these processes is being affected. The 1960–2020 total emissions (82% fossil and 18% land-use change) were partitioned among the atmosphere (47%), ocean (25%), and the land (30%), all responsible for the growth rate in atmospheric CO_2 concentration and in the changes in CO_2 sunk. However, at least a decade of continuous observations, monitoring, measurements, and data acquisition is needed to detect a sustained decrease of -1% in global emissions that is needed for a 66% likelihood of assessing the sinking behavior of the land and/or ocean [45].

This work will revise the published literature to identify the R&D gaps, needs and challenges that could make CO_2 utilization an enabler for a sustainable circular economy.

References

1. EPA Greenhouse gas emissions. Understanding global warming potentials. (2021) https://www.epa.gov/ghgemissions/understanding-global-warming-potentials. Accessed Feb 2022
2. IPCC Climate change and land. Special Report. Intergovernmental Panel on Climate Change (IPCC). (Cambridge University Press, UK, 2020), p. 540
3. M. Allen et al., Global warming of 1.5 °C. Special Report. Intergovernmental Panel on Climate Change (IPCC). (Cambridge University Press, UK, 2018), p. 540. https://www.ipcc.ch/sr15/
4. NGFS Network for greening the financial system: Scenarios portal. 2022. https://www.ngfs.net/ngfs-scenarios-portal/. Accessed Mar 2022
5. B.D. Hall et al., Revision of the world meteorological organization global atmosphere watch (wmo/gaw) CO2 calibration scale. Atmos. Meas. Tech. **14**(4), 3015–3032 (2021). https://doi.org/10.5194/amt-14-3015-2021
6. NOAA Trends in atmospheric carbon dioxide. (2022). https://gml.noaa.gov/ccgg/trends/. Accessed Feb 2022
7. M. Aresta, The carbon dioxide problem, in *An economy based on carbon dioxide and water, potential of large scale carbon dioxide utilization*, ed. by M.K. Aresta, Iftekhar, S. Kawi (Springer, Switzerland AG, 2019), pp. v–xi
8. Worldometer World population by year. 2021. https://www.worldometers.info/world-population/world-population-by-year/. Accessed Apr 2022
9. H. Ritchie et al., Energy production and consumption (2020), p. 15. https://ourworldindata.org/energy-production-consumption#citation. Accessed Apr 2022
10. H.-M. Zhang et al., NOAA global surface temperature dataset (noaaglobaltemp), version 5. (2022). https://psl.noaa.gov/cgi-bin/data/timeseries/timeseries1.pl. Accessed Apr 2022

11. X. Chen et al., Transition towards higher penetration of renewables: an overview of interlinked technical, environmental and socio-economic challenges. J. Mod. Power Syst. Clean Energy **7**(1), 1–8 (2019). https://doi.org/10.1007/s40565-018-0438-9
12. Global Carbon Project Global carbon atlas: CO2 emissions. 2019. http://www.globalcarbon atlas.org/en/CO2-emissions. Accessed on Feb 2021
13. International Energy Agency Global energy review: CO2 emissions in 2020. (IEA, Paris, France, 2020), p. 55. www.iea.org/articles/global-energy-review-co2-emissions-in-2020
14. EIA, Short-term energy outlook. (US Energy Information Administration: Washington, DC, 2019), p. 49. www.eia.gov/outlooks/steo/report/renew_co2.php
15. United Nations, United Nations framework convention on climate change. The Paris agreement. 2015 (Adopted Dec 12), p. 27. https://www.un.org/en/climatechange/paris-agreement. Accessed Dec 2021
16. United Nations, Glasgow climate pact. COP26: Decision-/CMA.3 (2021), p. 11. https://unfccc. int/sites/default/files/resource/cma3_auv_2_cover%2520decision.pdf. Accessed Dec 2021
17. P. Friedlingstein et al., Global carbon budget. Earth Syst. Sci. Data Discuss. **191** (2021) (Copernicus Publications). https://doi.org/10.5194/essd-2021-386
18. V. Masson-Delmotte et al., Climate change 2021: the physical science basis. Sixth Assessment Report. Intergovernmental Panel on Climate Change (IPCC). (Cambridge, UK, 2021), p. 1800
19. R. Moss, S. Schneider, Uncertainties in the IPCC tar: recommendations to lead authors for more consistent assessment and reporting, in *The Third Assessment Report of the IPCC: Guidance Papers on the Cross Cutting Issues*, ed. by R. Pachauri, et al. (World Meteorological Organization, Geneva, Switzerland, 2000), pp. 33–51
20. M.R. Allen et al., Describing scientific uncertainties in climate change to support analysis of risk and of options, in *Proceedings of IPCC Workshop Report*, (Boulder, CO, USA, 2004), pp. 53–57
21. National Research Council, *Advancing the Science of Climate Change*, (The National Academics Press, Washington, DC, 2010), p. 526. https://doi.org/10.17226/12782
22. T.F. Stocker et al., Climate change 2013: the physical science basis. Contribution of working group I to the fifth assessment report of the intergovernmental panel on climate change. IPCC. (Cambridge, United Kingdom and New York, NY, USA, 2013), p. 1535.
23. S. Dangendorf et al., Persistent acceleration in global sea-level rise since the 1960s. Nat. Clim. Chang. **9**(9), 705–710 (2019). https://doi.org/10.1038/s41558-019-0531-8
24. R.S. Nerem et al., Estimating mean sea level change from the topex and jason altimeter missions. Mar. Geodesy **33**, 435–446 (2010). https://doi.org/10.1080/01490419.2010.491031
25. B. Metz et al., Carbon dioxide capture and storage. Special Report. Intergovernmental Panel on Climate Change (IPCC) (Cambridge University Press, UK, 2005), p. 442
26. National Research Council, Energy and climate: Studies in geophysics, (The National Academies Press, Washington, DC, 1977), p. 174. https://www.nap.edu/catalog/12024/energy-and-climate-studies-in-geophysics. https://doi.org/10.17226/12024
27. C. Song et al., Cryogenic-based CO2 capture technologies: state-of-the-art developments and current challenges. Renew. Sustain. Energy Rev. **101**, 265–278 (2019). https://doi.org/10.1016/j.rser.2018.11.018
28. C. Song et al., Energy analysis of the cryogenic CO2 capture process based on stirling coolers. Energy **65**, 580–589 (2014). https://doi.org/10.1016/j.energy.2013.10.087
29. B. Li et al., Advances in CO2 capture technology: a patent review. Appl. Energy **102**, 1439–1447 (2013). https://doi.org/10.1016/j.apenergy.2012.09.009
30. P. Voser, The natural gas revolution. Energ. Strat. Rev. **1**(1), 3–4 (2012). https://doi.org/10.1016/j.esr.2011.12.001
31. R. Lindsey, Climate change: atmospheric carbon dioxide. (2021), p. 5. https://www.climate.gov/news-features/understanding-climate/climate-change-atmospheric-carbon-dioxide. Accessed Apr 2022
32. S. Holloway et al., A review of natural gas occurrences and releases and their relevance to CO2 storage. 2005/8, External Report BGS CR/05/104 Report. IEA Greenhouse gas R&D Programme; British Geological Survey. (2005), p. 117. https://ieaghg.org/docs/General_Docs/Reports/2005-8.pdf

33. Y. Xiao et al., Natural CO2 occurrence in geological formations and the implications on CO2 storage capacity and site selection. Proc. Energy Procedia **4**, 4688–4695 (2011) (Elsevier Ltd.). https://doi.org/10.1016/j.egypro.2011.02.430
34. M. Peters et al., CO2: from waste to value. Chem. Eng. **813**, 46–47 (2009)
35. J. Klankermayer, W. Leitner, Harnessing renewable energy with CO2 for the chemical value chain: challenges and opportunities for catalysis. Philos. Trans. R. Soc. A: Math., Phys. Eng. Sci. **374**(2061) (2016). https://doi.org/10.1098/rsta.2015.0315
36. J. Klankermayer et al., Selective catalytic synthesis using the combination of carbon dioxide and hydrogen: catalytic chess at the interface of energy and chemistry. Angew. Chem. Int. Ed. **55**(26), 7296–7343 (2016). https://doi.org/10.1002/anie.201507458
37. T.E. Müller, W. Leitner, CO2 chemistry. Beilstein J. Org. Chem. **11**, 675–677 (2015). https://doi.org/10.3762/bjoc.11.76
38. B.A. Kimball et al., Effects of increasing atmospheric CO2 on vegetation. Vegetatio **104–105**(1), 65–75 (1993). https://doi.org/10.1007/BF00048145
39. S.B. Idso et al., Effects of atmospheric CO2 enrichment on plant growth: the interactive role of air temperature. Agr. Ecosyst. Environ. **20**(1), 1–10 (1987). https://doi.org/10.1016/0167-8809(87)90023-5
40. E. Rosenthal, Cannabis and CO2: Why plants suffer if they don't get at least a minimum amount of CO2. Guru of Granja Blog (2020), p. 11. https://www.edrosenthal.com/the-guru-of-ganja-blog/why-co2-is-critical-for-cannabis. Accessed Apr 2022
41. F. Wang et al., Higher atmospheric CO2 levels favor C3 plants over C4 plants in utilizing ammonium as a nitrogen source. Front. Plant Sci. **11**(2020). https://doi.org/10.3389/fpls.2020.537443
42. J. Lovelock, A new look at life on earth, in *The Ages of Gaia: A Biography of Our Living Earth*. ed. by L. James (Oxford University Press, Oxford, UK, 1988), pp. 135–139
43. J.E. Lovelock, M. Whitfield, Life span of the biosphere. Nature **296**(5857), 561–563 (1982). https://doi.org/10.1038/296561a0
44. C.D. Campbell et al., Estimation of the whole-plant CO2 compensation point of tobaco (nicotiana tabacum L.). Glob. Chang. Biol. **11**(11) (2005), 1956–1967. https://doi.org/10.1111/j.1365-2486.2005.01045.x
45. G.P. Peters et al., Towards real-time verification of CO2 emissions. Nat. Clim. Chang. **7**(12), 848–850 (2017). https://doi.org/10.1038/s41558-017-0013-9

Chapter 2
Decarbonization

2.1 Introduction

Some chemical processes emit significant amounts of greenhouse gases (e.g., methane, carbon dioxide, etc.) since the required thermal energy is generally provided by burning fossil fuels (natural gas, crude oil, and coal). However, as mentioned above, the low efficiency (27–35%) of processes converting chemical energy into thermal, mechanical, or electrical energy leads to a direct release of heat into the atmosphere of the associated 65–73% ineffective conversion.

CO_2 emissions link energy to climate change since the Energy Sector is responsible for more than 60% of the global CO_2 emissions that increases to 75% when the rest of the industrial sector is included. Within the industrial processes, for the cement, steel, ethylene, and ammonia plants, 90% of their GHG emissions are CO_2, representing over 23% of the total global emissions [1, 2]. Improvements in the efficiency of the current energy systems must be part of any approach or strategy leading to the mitigation of climate change. Clearly, both emissions and heat should be recovered and used or recycled.

This situation has reached risky limits for the survival of the life on Earth, which is why it is necessary to develop strategies that contribute positively to reduce the emission of CO_2 into the environment and to find alternatives to satisfy our fuel needs. This scenario suggests a radical change of the energy vector in which more efficient, clean, and secure energy sources should be considered, and this points towards energy decarbonization as the key topic of R&D activities. Therefore, it is necessary to make technological and economic efforts worldwide to achieve this goal by decarbonizing fuel energy sources. Additionally, governments are defining policies, mechanisms, and investment plans that enable the strategy execution. Pricing carbon emissions is one of the subjects considered in policy definitions. Two instruments for carbon pricing have been, are being implemented or are being defined, to achieve the required acceleration on the reduction of CO_2 emissions, namely, emissions trading systems and carbon taxes.

© The Author(s), under exclusive license to Springer Nature Switzerland AG 2023
M. M. Ramirez-Corredores et al., *Decarbonization as a Route Towards Sustainable Circularity*, SpringerBriefs in Applied Sciences and Technology, https://doi.org/10.1007/978-3-031-19999-8_2

Decarbonization implies the reduction of environmental carbon. Currently, the reduction of CO_2 emissions represents the main target in many decarbonizing strategies, especially from fossil-fuel-fired power plants, which are the largest source of CO_2 emissions, accounting for roughly 40% of total CO_2 emissions [3]. In order to achieve decarbonization, it is necessary to develop new technologies [4] that.

(i)   capture and store or sequester the produced CO_2 (CS),
(ii)   convert the produced CO_2 into more valuable chemical feedstock (CU), carbon avoidance through:
(iii) recycle the produced CO_2 and/or
(iv)   eliminate the use of any process that produce CO_2 emissions.

"Carbon sequestration" mentioned above is a process for capturing (CC) and storing (CS) emitted or atmospheric CO_2, in a stable state, at long-term scale. Biological, chemical, geological, or physical processes are known. The term "carbon fixation" is used to describe the sequestration process carried out by natural ecosystems (forests and oceans) and in these cases the uptake mechanisms are referred to as "carbon sinks".

A decarbonizing strategy, regardless of any accelerated mechanism, is not a straightforward path. Instead, it involves a transition stage in which CCUS technologies will play key roles in mitigating global warming during this period. Additionally, incremental incorporation of utilization technologies will alleviate the needs for storing or sequestering processes. As exemplified in Fig. 2.1, the considered and investigated CCUS concepts would enable a transition from the present gray towards a green future, hopefully without an extensive need for brown (also called blue) processes.

Clearly, processes for the conversion of CO_2 (using low-carbon energy sources) are needed to warrant that green future. Moreover, integration of these processes

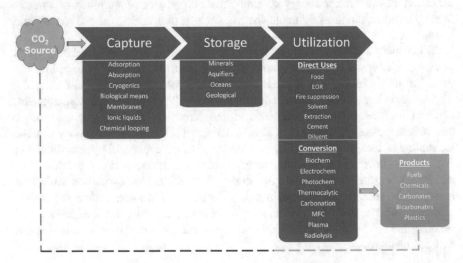

Fig. 2.1 CCUS conceptual basis and role of conversion in a "green future"

within circular loops must be a primary objective. Diverse R&D activities carried out at universities and research centers are considering CO_2 as a valuable source of carbon with topics related to CO_2 reuse from different perspectives, for instance, strategic, industrial, scientific, and social, with arguments highly interconnecting all these aspects. Capture and conversion of CO_2 to high value-added products could provide new routes within an environmentally sustainable economy [5]. Some of these processes are discussed below.

Natural CO_2 fixation occurs via the photosynthesis process that transforms CO_2 and water into different types of carbohydrates, using solar energy. Also, photosynthesis is balanced by the oxidation of carbon compounds back to carbon dioxide and water through respiration, decay, and combustion. This dynamic cycle is able to maintain the concentration of carbon dioxide in the atmosphere, thereby sustaining life on earth. However, the large use of fossil fuels is unbalancing this situation, bringing about increasing concentrations of carbon dioxide in the atmosphere, which may eventually result in global warming via the well-known greenhouse effect.

Although natural photosynthesis will always play an important role in carbon fixation, CO_2 abatement cannot rely on natural photosynthesis as the unique process to reduce CO_2 concentration in the atmosphere. Hence, alternative ways of CO_2 fixation must be searched. Among them, one may highlight routes using CO_2 as a reactant to produce value-added chemicals, fuels, or materials.

Regardless of all the advances and efforts on CCS R&D, the risks (specially leakage) and economy constitute the causes for deployment struggle. The sequestration difficulties to formulate a value proposition are closely followed by the R&D and industrial communities who are turning their attention towards utilization, in the search for economic incentives for an approach to pursue climate change mitigation. Some of the physical and chemical approaches shown in Fig. 2.1 have been reviewed [6]. Although physical processes are mature technologies, their demand is quite low and such pathways need to rely on market dynamics, which in turn depend on policies and global and/or local energy prices. Meanwhile, broader opportunities were identified for the chemical pathways creating a huge demand for R&D in all areas concerning process technologies development (e.g., units design, catalysis, reactor and process configuration, techno-economic analysis (TEA), life cycle assessment (LCA), etc.).

CO_2 utilization will be one of the main focuses of this book. However, one may already anticipate that CO_2 has several industrial uses and is currently being used in different applications in almost all economic sectors. In fact, it has found applications in the food industry, the oil industry, and the chemical industry. While the largest volume is used by the oil industry, another of its greatest uses is as a chemical, in the food industry, or rather, in the production of carbonated beverages. CO_2 provides the sparkle in carbonated beverages such as soda water, beer, and sparkling wine. Also, CO_2 is a food additive used as a propellant and acidity regulator in the food industry. In addition, CO_2 may be utilized as leavening agent, which causes dough to rise by producing CO_2 back. In fact, baking powder and baking soda release CO_2 when heated or if exposed to acids.

Being an inert gas, CO_2 is used in fire extinguishers since it extinguishes flames by flooding the environment around the flame. In fact, some fire extinguishers, especially those designed for electrical fires, contain liquid carbon dioxide under pressure [7].

Another interesting CO_2 application concerns its use as a supercritical solvent. Indeed, CO_2 has attracted attention in the pharmaceutical and other chemical-processing industries as a less toxic alternative to more traditional solvents such as organochlorides. An emerging separation technology, the Supercritical Fluid Extraction, employs compounds such as CO_2, at conditions above their critical points as an organic solvent alternative, due to its low cost, low reactivity (inert gas), low toxicity, and mild critical conditions [8–10].

CO_2 may also find application in the form of dry ice, which is often used during the cold soak phase in winemaking to cool clusters of grapes quickly after picking. The use of dry ice prevents spontaneous fermentation by wild yeast in the wine production [11].

Other applications that may be highlighted are the medical and pharmacological uses, enhanced oil recovery via CO_2 injection in wells, and transformation into fuel via algae growth which could then be converted into biodiesel fuel. However, it must be borne in mind that not all aforementioned applications represent CO_2 fixation. Although these latter applications do not valorize the CO_2, they contribute to its recycling and reuse, which could be part of integrated circular business models that will be considered and discussed in Chap. 3. Currently, industry uses less than 250 Mton/y of CO_2, which represent less than 0.7% of the global emissions. In fact, reported industrial used volume was 230 Mton/y, by 2018, with a major utilization from the oil and gas sector distributed in 130 Mton/y for urea manufacturing and 70–80 Mton/y for enhanced oil recovery, EOR [12]. The following sections of this chapter will review and discuss the main chemical fixation routes categorized as CO_2 valorization via inorganic routes, organic routes, carboxylation, and artificial photosynthesis. Thus, a more detailed discussion follows.

2.2 CO_2 Inorganic Valorization

Formation of inorganic salts has been one of the bases for C-storage and sequestration. CO_2 mitigation by means of the formation of carbonates usually concerns the concept of geological CO_2 capture and storage. Indeed, such fixation step comprises injections of carbon dioxide into mature or depleted reservoirs and deep saline aquifers. Carbon dioxide undergoes reaction with basic ions, yielding carbonates [13, 14]. CCS technologies are well recognized as the only commercially available processes, capable of providing significant emission reductions from fossil-fuel-fired power plants and from industrial manufacturing sector. The environmental sustainability of CCS projects requires a proper selection, characterization, and management of the site and its operation, which becomes long-term economic intense that mostly jeopardize economic feasibility. This lack of economic sustainability has called for value-adding utilization applications of the captured and/or stored CO_2. Small standalone emitters located remotely from adequate geological storage sites cannot benefit from geologic sequestration. One of the most benefited areas of application of these principles has been Enhanced Oil Recovery (EOR), which currently represents one of the largest volumes of CCS-CO_2 utilization. However, CCS falls outside the scope of the present work and the reader is directed to the many books and reviews published in this area (e.g., Refs. [15–21]).

2.2.1 Carbonation

Mineralization or mineral carbonation (MC) refers to the process of capturing CO_2 into alkaline or alkaline earth carbonates, bicarbonates, or carbamates. MC refers to conversion processes mimicking some of the natural fixation reactions occurring between atmospheric gases and earth surface minerals. Thus, in practice, mineralization can be carried out either ex situ through a chemical-capturing process or in situ, e.g., geological formations, for sequestration purposes. The topic has been the subject of books [22–24] and reviews [25–31] where the reader can find or search for more detailed information. The acidic solution of CO_2-water promotes the dissolution of silicate minerals (e.g., olivine, wollastonite, and serpentine), providing metal (Me) cations that can react with the dissolved CO_2, forming stable carbonates and consuming protons, so the pH increases and the carbonate precipitates as neutralization advances (see reactions 2.1 and 2.2). The reaction kinetics is controlled by the cations release from the silicate mineral, which depends on the type of mineral, element, pH, and temperature. In fact, the high reactivity and abundance of divalent metal cations contained in silicates are responsible for ~30% of the natural reduction of atmospheric CO_2. Silicates are an intrinsic part of basaltic rocks that are the most abundant of the ocean floor, ~70% of the Earth's surface and more than 5% of the continents [32]. The most important basalt fields are located in India, Siberia, the United States, Canada, and Yemen.

$$2H^+ + H_2O + (Ca, Mg, Fe)SiO_3 \rightarrow Ca^{2+}, Mg^{2+}, Fe^{2+} + H_4SiO_4 \qquad (2.1)$$

$$H_2CO_3 + Me^{2+} \rightarrow MeCO_3 + 2H^+ \qquad (2.2)$$

Although thermodynamically favored as an exothermic process (see reactions 2.3 through 2.5), MC is kinetically limited (very slow). The energy intensity of the processes is not alleviated by the exothermicity of the carbonation reaction. Energy consumption is due to treatments of wastes and mineral rocks, to the recovery of unconverted reactants, and to the reaction rate acceleration induced by chemical additions or process conditions. Although the acceleration of the kinetics has been proposed to be sorted out, in many ways, neither the assessment of the energy intensity nor lifecycle analyses have been reported. Therefore, the determination of the CO_2 produced per ton of stored CO_2 remains unknown.

$$Mg_2SiO_4 + 2CO_2 \rightarrow 2MgCO_3 + SiO_2 \qquad \Delta H_{298K} = -89kJ/mol \qquad (2.3)$$

$$Mg_3Si_2O_5(OH)_4 + 3CO_2 \rightarrow 3MgCO_3 + 2SiO_2 + 2H_2O \quad \Delta H_{298K} = -64kJ/mol \qquad (2.4)$$

$$Ca_2SiO_4 + 2CO_2 \rightarrow 2CaCO_3 + SiO_2 \qquad \Delta H_{298k} = -90kJ/mol \qquad (2.5)$$

Compared to CO_2, carbonates exhibit a lower energy state, releasing substantial amounts of energy during their formation which occur spontaneously in nature, nonetheless very slowly. Since carbonation yields stable solid mineral products, it could be used for the direct sequestration of CO_2 from flue gas, if the kinetic restrictions are overcome [33]. For this purpose, the slow reaction rate of silicate rocks requires mechanical, chemical, or biological activation. Additionally, material balance (stoichiometry) implies a larger amount of mineral or wastes to be handled per unit of CO_2 to be removed (2–3 ton$_{rock}$/ton$_{CO2}$). Finally, the aggregated value of the products is low, probably due to the supply/demand market balance. Both scientific research and industrial development have focused their activities on the acceleration of the kinetics, using cost-effective materials with minimal environmental impact. Many minerals have been used aiming at CO_2 sequestration via carbonation. There are several companies working on the carbonation of Mg-rich sources from different perspectives like the carbonation of natural Mg-silicates or of Mg(hydro)oxides. For instance, the application of a staged process for CO_2 sequestration by mineralization using serpentinite and metaperidotite (types of magnesium silicates) has been proposed [34]. Mg can be extracted from such minerals using recoverable ammonium salts (ammonium sulfate—AS and ammonium bisulfate—ABS), precipitated in the form of $Mg(OH)_2$ and subsequently carbonated at 20 bar CO_2 partial pressure. Such chemical route seems rather promising. Indeed, the inherent exothermicity of the MC reactions can result in the total process energy balance attaining neutral status [35].

Regardless of these potentialities, additional limitations have been identified such as the kinetics limitations, the size of the produced salt market, the process scale and logistics, the energy intensity, and the realization of the direct capture from flue gases. The size of the market is proportional to the uses that the formed carbonates currently have. Table 2.1 provides a brief description of the current uses of the main mineralization-derived carbonates. Pioneer efforts in the development of ex situ mineralization, employing calcium and/or magnesium silicates started at Los Alamos National Laboratory, by the second half of the 1990's decade [36]. The evolving work carried out at the National Energy Technology Laboratory (NETL) led to the direct aqueous mineralization method [37–39]. Concerning aqueous carbonation, it is a process in which high-pressure CO_2 is injected into water or a sodium bicarbonate solution mixed with olivine or serpentine in order to produce magnesium carbonate [37]. Thus, CO_2 gas is fixed in a solid carbonate. The process comprises two basic steps: magnesium ion dissolution from the olivine or serpentine and magnesite precipitation [40]. Nevertheless, such reactions are generally slow, and much

Table 2.1 Current uses and applications of carbonates

Carbonate	Uses
Alkaline	
Lithium carbonate	Drug development
Sodium carbonate (soda ash)	Glass making, pulp and paper industry, sodium chemicals (silicates), soap and detergent production, paper industry, and water softener
Potassium carbonate	Glass making, soft soap production, textile, and photography chemicals
Rubidium carbonate	Glass making, short-chain alcohol production
Cesium carbonate	Production of other cesium salts
Alkaline Earth	
Beryllium carbonate	Processing ores and in chemical and nuclear applications
Magnesium carbonate	Skin care products, cosmetic, anti-fire products, climbing chalk
Calcium carbonate	Glass, textile, paint, paper and plastic production, caulks industry, to produce ink and sealant, non-toxic food additive, as a drug development, and chalk production
Strontium carbonate	Fireworks, magnets, and ceramic manufacture
Barium carbonate	Glass, cement, ceramic, porcelain, rat poison manufacture
Boron family	
Aluminum carbonate	Drug development
Thallium carbonate	Fungicides manufacture
Others	
Lead carbonate	Glass, cement, ceramic, porcelain, and rat poison manufacture

research has been carried out to increase the reaction rate. Although costs were at first somewhat high, process improvements brought about a considerable reduction in such costs. Notwithstanding, the scale of ex situ MC operations, requiring approximately 55,000 ton of mineral to carbonate, the daily CO$_2$ emissions from a 1-GW, coal-fired power plant may render such operations impracticable [39]. Although mineralization at small scales is commercially deployed for niche applications of CO$_2$ sequestration, deriving revenues from by-products, larger scale applications are slowing down due to the limited economic feasibility. Worldwide progresses have been made and applications and utilization pathways have been investigated, some of them advancing to larger scales (Fig. 2.2).

Figure 2.2 shows that valuable products (e.g., metals, cements, and construction materials), useful by-products (e.g., silica, chemicals) as well as productive applications (e.g., remediation of negatively valued waste feedstocks) could serve as the basis for business cases to make mineralization feasible, even in the absence of a carbon price. Additionally, policies and policy mechanisms are also required not only to price CO$_2$ sequestration independently of emission reductions, but also to promote an intrinsic value of carbon. Both supporting and enabling policies are

Fig. 2.2 Mineralization
pathways

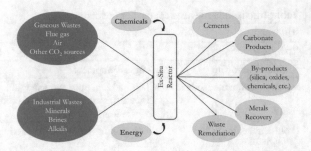

needed. However, it must be borne in mind that, for a policy to be effective, governments ought to recognize the importance of CCU, not only environmental aspects but also economic ones. Governments must see CO_2 as a commodity chemical feedstock and not as a waste. The idea of considering CO_2 as a commodity will generate products that have market value and so confer profitability to the economy [41].

Process economy is also affected by logistic issues, such as the emissions sites and production of silicates, as well as the location of carbonate end-users. Finland is a good example of logistic issues. In fact, carbon dioxide mineralization was identified as the only option for CCS (carbon dioxide capture and storage) application in Finland. Unfortunately, such solution has not been adopted by the power sector, probably because the most suitable mineral resources are found in central and northern Finland whereas most fossil-fuel-fired electricity production is located in southern Finland. Nevertheless, one interesting source-sink combination is the magnesium silicate occurrences at Vammala, located about 85 km east of the 565 MWe coal-fired Meri-Pori power plant on the country's southwest coast, which produces 2.5 $Mton_{CO2}$/y. Fortum and TVO enterprises have studied the deployment of CCS for a potential retrofitting of the Meri-Pori power plant. However, the project was abandoned. Notwithstanding, an interesting chemical route was developed based upon the Meri-Pori case. Experimental work indicated that carbon dioxide mineralization without CO_2 pre-capture could be directly applied to flue gases containing sulfur oxides. In the study performed, carbonation of $Mg(OH)_2$ with CO_2 was compared with CO_2-SO_2-O_2 gas mixtures. Interestingly, results demonstrate that SO_2 promptly reacts with $Mg(OH)_2$, thereby enabling one to simultaneously capture both SO_2 and CO_2. Hence, the flue gas desulfurization step would become unnecessary [42]. Unfortunately, the development of this process was abandoned due to the high price of NaOH and the difficulty in recycling it [43]. Both energy and effectiveness of the raw material are additional drawbacks of the process, including a total penalty of about 3GJ (mainly derived from the 400 °C heat) and the usage of 3 ton_{rock}/1000 kg_{CO2} [44]. This type of process underwent several modifications. In fact, aiming at achieving higher purity and deliver more valuable mineral products, the process was divided into four steps. The first step consists of Mg leaching from the magnesium silicate using HCl, followed by a three-step precipitation in consecutive reactors to remove $Fe(OH)_3$, then $Fe(OH)_2$ and other divalent ions, and finally $MgCO_3$ nucleation and growth. The technical viability of the process was proven [45]. However, economic issues of the process must consider the cost of raw materials such as serpentinite. Moreover, an energy-neutral process depends on the availability and quality of nearby waste heat. Also, economic viability demands high magnesium extraction and carbonation levels. Eventually, one must consider the processing of CO$_2$-containing flue gases (eliminating the expensive capture step) and production of marketable products [46, 47]. R&D on new applications and uses of the resulting products could benefit economy as well. For instance, the UK has a readily accessible serpentine reserve of 2–10 Gton located only 2 km away from the coast, enabling cost-effective mass transport links. The current high cost of carbon capture and storage by mineralization (CCSM) may decrease by direct flue gas capture, process optimization, and system integration at each site. CCSM has the potential to become an economically viable alternative to capture and store CO$_2$ from small and medium emitters [48].

The commercial potential for mineralization using magnesium serpentines has been reported by Cambridge Carbon Capture (CCC) Ltd., an UK based company [49]. They have found out that 6 billion tons of serpentine resource from the largest magnesium silicate mine in the world could satisfy the current market demand for serpentine (~20 Mton/y) and provide a significant volume of the required raw material for mineralization. The carbonation process developed by this company (CCC) is based on the alkaline digestion of silicate minerals and CO$_2$ wastes [50, 51]. The activating pretreatment of the silicate minerals (olivine, serpentine) was also developed to produce a more reactive material (brucite or magnesia) [52]. This type of process is not free from by-products and wastes (e.g., silica, metals, magnesia, carbonate). However, metals production could lead to economic benefits [53].

Co-production of electricity via fuel cell has been suggested as a profit generation mechanism. CCC made announcements of having started the development of such technology, though, so far, no results have been published on this regard [54].

The reactions' slowness calls for accelerating and energy-intensive treatments, such as the use of chemical additives or more severe process conditions must be considered. In both cases, either increased emissions or increased waste generation are subjects for attention. Therefore, the mass of generated CO$_2$ per ton of stored CO$_2$ through a detailed Life Cycle Assessment (LCA) needs to be determined [55]. Attempts to accelerate the kinetics includes pretreatment and activation processes of minerals, type of raw material used (mineral or residue), reacting phases (gas–solid or aqueous), and source of CO$_2$ (pure CO$_2$ or flue gas) [28].

A comparison of two MC routes producing either magnesite (dry pathway) or (hydro)magnesites (wet pathway) operating directly on flue gas showed similar conversion levels and rates for the two routes, though slightly better for the wet route. As a matter of fact, the dry carbonation route with lime kiln gas has the lowest exergy demand of the processes considered (2.6 GJ heat and 0.9 GJ electricity per ton CO$_2$). Therefore, it is remarkably better than the wet process route (15.4 GJ heat per ton CO$_2$). On the other hand, the wet route has a zero-exergy input as power, requiring only (waste) heat. For both lime kiln routes, the heat required can be obtained from the kiln gas; notwithstanding, the wet process route does not require flue gas compression. When flue gas from a natural-gas-fired power plant must be treated, the exergy demand on the CCUS plant is significantly larger than the lime kiln (11.4 GJ heat and 8 GJ electricity per ton CO_2 for a dry process, or 12.4 GJ heat for the wet process). This occurs due to the much larger amounts of gas being treated. Eventually it is important to remember that in the case of the wet carbonation process route, excess NH_3 may be needed to establish the working pH for hydromagnesite precipitation. Hence, this route presents a potential risk of excessive NH_3 losses which must be prevented [56]. The needs for intensification are clear. For instance, intensification of carbonation by the direct reaction with flue gases would enable sequestration avoiding investments on energy-intensive large-scale solvent capture and overcome public acceptability challenges for investing on transport and storage infrastructure of supercritical CO_2 [57].

Once more R&D is needed to develop new processes, likely based on new carbonation chemistries and better process engineering schemes, for minimizing the net lifecycle energy per ton of stored CO_2 as carbonate. Emphasis should be placed on including investigation on the most energy-intense parts of the process.

An electrochemical approach to produce hydrogen during the carbonation process has been proposed. Such process involves the construction of an innovative electrochemical cell comprising a dual-component graphite and aluminum anode, a hydrogen-producing cathode, and an aqueous sodium chloride electrolyte. An anodic CO_2 electrochemical process aims at transforming CO_2 into a safe molecule capable of being stored. Bicarbonate is formed and reacts with anodic metal generated cations, yielding mineralized carbon dioxide. Interestingly, whereas conventional electrochemical carbon dioxide reduction requires hydrogen, this cell produces hydrogen at the cathode. At the anode, the cell displays both oxidative and capacitive electrochemistry. In order to promote carbon capture in a very sustainable way, scrap metal may be used within the anode. Also, seawater may be the electrolyte, whereas an industrial gas stream and a solar panel are deployed as a zero-carbon energy source [58]. After quantifying the H_2 cathode by-product, results showed that such process might be a net energy producer and recycled aluminum should be used as the sacrificial anode [59].

Production of ammonium bicarbonate, using CO_2 as co-reactant with ammonia (NH_3) and water, represents an interesting route of chemical sequestration. Ammonium bicarbonate can be produced by combining carbon dioxide and ammonia according to the reaction shown as 2.6.

$$CO_2 + NH_3 + H_2O \rightarrow (NH_4)HCO_3 \tag{2.6}$$

In other to avoid thermal decomposition and facilitate the precipitation of the thermally unstable ammonium bicarbonate, the reaction solution is kept cold, and the product is then recovered as a white solid. Ammonium bicarbonate is commonly used as an inexpensive nitrogen fertilizer, if used in sugarcane crops. Its production is facilitated using the pure CO_2 from ethanol-producing plants, based on sugarcane juice fermentation. This recycling of CO_2 from ethanol-producing plants into the fertilizer for harvesting the sugarcane crops is a good example of circular economy [60].

Natural carbonation of silicate rocks is very slow. Hence, the addition of a mineral dissolution step using chemicals is an efficient technique to shorten reaction times and enhance the reaction extent. Among other chemical routes, MC using recyclable ammonium salt's pH swing processes is considered a very promising MC technique to store CO$_2$ permanently. The process comprises five steps: (1) CO$_2$ capture using ammonia (NH$_3$ + CO$_2$ $\leftrightarrow$ NH$_4$HCO$_3$); (2) the leaching of the Mg/Ca cations from the mineral resource using acid ammonium bisulfate solution (NH$_4$HSO$_4$ + Mg/Ca-rich silicate $\leftrightarrow$ MgSO$_4$ + SiO$_2$ + unreacted silicate + (NH$_4$)$_2$SO$_4$); (3) pH-regulation (to swing the pH from pH 1–2 caused by unreacted NH$_4$HSO$_4$, to pH 8–9, at which carbonation reaction occurs); (4) the MC of CO$_2$ (as NH$_4$HCO$_3$) (MgSO$_4$ + NH$_4$HCO$_3$ + H$_2$O $\leftrightarrow$ MgCO$_3$·3H$_2$O + (NH$_4$)$_2$SO$_4$ + CO$_2$); and (5) the regeneration of the used chemicals. However, the main key challenge of this process at large scale concerns the energy consumption associated to the regeneration of the employed additives and specially to the separation of the salt to be regenerated from the water solution. Liquid–liquid extraction to replace the energy-intensive salts/water separation step has been proposed [61], resulting in an energy saving of 35% in comparison to water evaporation. Furthermore, the process became carbon negative when water evaporation was replaced by extraction.

2.2.2 Urea Synthesis

Urea is an amide being widely used in fertilizers as a source of nitrogen (N); also, it is an important raw material for the chemical industry. Besides its use as fertilizer, urea is a raw material for the manufacture of two main classes of materials: urea-formaldehyde resins and urea-melamine-formaldehyde [62]. Furthermore, urea may be used as a co-reactant, in processes like the Selective Non-Catalytic Reduction [63] and the Selective Catalytic Reduction [64], to reduce the NOx pollutants in exhaust gases from combustion of diesel, of dual fuel, and of lean-burn natural gas engines. In the health area, urea-containing creams are used as topical dermatological products to promote rehydration of the skin [65].

Urea is industrially produced from NH$_3$ and CO$_2$ (2.7 and 2.8). The synthesis of urea is currently the largest CO$_2$ conversion process by volume in the industry. Although urea is an organic compound, the synthesis thereof involves inorganic molecules, namely, CO$_2$ and NH$_3$ [66]. Urea production plants are almost always located adjacent to the site where the ammonia is manufactured, since both ammonia and CO$_2$ are available in the site. The industrial plant comprises two main equilibrium reactions, with incomplete conversion of the reactants. The first step is the carbamate formation, an exothermic reaction of liquid ammonia with gaseous CO$_2$ (2.7) at high temperature and pressure to form ammonium carbamate (H$_2$N-COONH$_4$). Next, the urea conversion takes place. This is an endothermic decomposition of ammonium carbamate into urea and water (2.8). Ammonia is generally produced using natural gas, which undergoes the well-known steam reforming reaction (2.9) followed by the water–gas shift (WGS) reaction (2.10) to yield H$_2$ and CO$_2$.

$$2NH_3 + CO_2 \rightleftharpoons H_2N - COONH_4 \quad \Delta H_{298K} = -117kJ/mol \quad (2.7)$$

$$H_2N - COONH4 \rightleftharpoons (NH_2)_2CO + H_2O \quad \Delta H_{298k} = 15.5kJ/mol \quad (2.8)$$

$$CH_4 + H_2O \rightarrow CO + 3H_2 \quad \Delta H_{298k} = 206kJ/mol \quad (2.9)$$

$$CO + H_2O \rightarrow CO_2 + H_2 \quad \Delta H_{298K} = -41.15kJ/mol \quad (2.10)$$

Recently, efforts for developing a greener urea-producing process have been undertaken, using wastes as feedstock. For instance, recovery of ammonia from sewage and livestock manure, a major environmental pollutant, has been studied in pursue of urea production [67].

2.3 CO_2 Organic Valorization

Production of a wide variety of organic compounds offers a myriad of opportunities for the valorization of CO_2. Particularly, fuels and chemicals are of interest due to the potential for major utilization of the end-products, by the end-users, additionally to the massive consumption of CO_2 volumes in their manufacture. The conversion of CO_2 is one step within the carbon cycle. Starting from CO_2, intermediates, building blocks, and final products can be obtained through a variety of chemistries. Nevertheless, attention should be placed to energy efficiency and low or neutral carbon energy sources. In 2015, a set of 12 principles, based on the acronym CO_2 CHEMISTRY, was proposed to structure the basis of any feasibility assessment of processes or reactions, which convert CO_2 into organic chemicals. These principles [68] are collected below:

C	Catalysis is crucial
O	Origin of the CO_2?
2	Tomorrow's world may be different
C	Cleaner than existing process?
H	High-volume or high-value products?
E	E-factor must be low (waste generated per unit weight of product)
M	Maximize integration
I	Innovative process technology
S	Sustainability is essential
T	Thermodynamics cannot be beaten
R	Renewable (and reasonable) energy input
Y	Your enthusiasm is not enough

2.3.1 Synthesis Gas (Syngas) Production

Syngas is the name given to a mixture of carbon monoxide (CO) and hydrogen (H_2) since it is a useful intermediate, from which a wide range of chemicals, fuels, and products can be synthesized (see Fig. 2.3). Among the products, olefins and methanol are of more significant importance as these are also intermediates and/or building blocks for some other products.

The global syngas production through 2019 had an associated consumption of 141.45 exaJ of natural gas, 193.03 exaJ of crude oil, and 157.86 exaJ of coal [69]. Syngas is usually produced via steam reforming of hydrocarbons (see reaction 2.9, for the case of methane steam reforming, SMR). Due to market fluctuations and requirements, several process technologies are commercially available to capture the high potentiality and versatility of syngas and satisfy market needs. Unfortunately, all these processes consume carbon from fossil resources resulting in high intensity for both carbon and energy.

2.3.1.1 CO₂ Reforming of Methane

Among the multiple possible reactions that CO_2 can undergo, the CO_2 reforming of methane, also known as methane dry reforming (MDR, reaction 2.11), produces synthesis gas (CO + H_2) with a molar H_2/CO of 1 [70, 71]. There are abundant reserves of natural gas with significant proportions of CO_2, which can serve as raw material to the process of MDR without the need to carry out costly gas separation or additional processes, including renewable sources such as biogas.

$$CO_2 + CH_4 \rightarrow 2CO + 2H_2 \qquad \Delta H_{298K} = 247 kJ/mol \qquad (2.11)$$

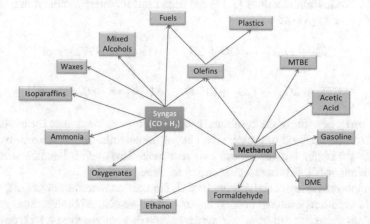

Fig. 2.3 Syngas products chain

The MDR has a more significant environmental importance since both reactants are the largest contributors to global warming. The efforts and attention from the scientific and industrial community have been growing in recent times. In addition, the H_2/CO ratio of the syngas produced by this process is a more convenient ratio, for further applications. In this regard, a higher H_2/CO ratio favors methane and inhibits chain growth [72–74], while a lower H_2/CO ratio favors the methanol synthesis, enabling the production of gasoline and other hydrocarbon families (paraffins, olefins, and aromatics) [75]. Alkenes, oxygenates, hydroformylation, and acetic acid production are also favored at low H_2/CO ratios [76].

Due to the high stability of the CO_2 molecule, its activation to be used as a methane oxidant in the MDR requires high energy, making it the most endothermic methane reforming reaction [77]. Several limitations have precluded MDR to reach commercialization, including, but not limited to, a large energy consumption due to the high endothermicity of the reaction, the extended deactivation of the employed catalysts by coke deposition, and the decrease of the H_2/CO ratio due to the inevitable formation of water as a by-product. The occurrence of the undesired reverse water–gas shift reaction (RWGS, 2.12) is one of the main sources of water formation that consumes part of the produced hydrogen and forms additional CO, resulting in H_2/CO ratios below unity [78–80].

$$CO_2 + H_2 \rightarrow CO + H_2O \quad \Delta H_{298K} = 41.15 kJ/mol \quad (2.12)$$

The high thermodynamic potential ($\Delta G^{\circ}_{298} = 174.6$ kJ mol^{-1}) is another major problem of MDR, which drives the use of high temperatures to run under a diminished free energy, e.g., $\Delta G^{\circ}_{1073} = -44.76$ kJ mol^{-1} [81, 82]. However, it has been determined that adding small amounts of oxygen or water significantly reduced carbon formation, minimized loss in syngas production, and reduced energy requirements. A combination of MDR with the exothermic methane partial oxidation (MPO, 2.13) and/or total oxidation reactions (2.14) has been used for energy intensification, with promising results [83–85].

$$CH_4 + 1/2O_2 \rightarrow CO + 2H_2 \quad \Delta H_{298k} = -36 kJ/mol \quad (2.13)$$

$$CH_4 + 2O_2 \rightarrow CO_2 + 2H_2O \quad \Delta H_{298k} = -802 kJ/mol \quad (2.14)$$

The energy benefits of the combined process become evident since the exothermic reaction can be coupled to provide the heat of the endothermic reactions, by facilitating heat transfer between these two reactions. The overall balance is a more energy-efficient MDR process, consuming less energy [86].

The important role of catalysts on the CO_2 conversion reactions is jeopardized by the use of high temperatures needed in view of the associated high thermodynamic potential. At these temperatures, carbon deposition through reaction 2.15 is promoted and deactivation takes place. Meanwhile, the exothermic Boudouard reaction (i.e., the carbon monoxide disproportionation shown as 2.16 reaction) is favored at lower temperatures [87].

$$CH_4 \rightarrow C + 2H_2 \quad \Delta H_{298K} = 74.85 kJ/mol \quad (2.15)$$

$$2CO \rightarrow C + CO_2 \quad \Delta H_{298k} = -172 kJ/mol \quad (2.16)$$

In addition to carbon formation on the active catalyst surface, the required high operating temperatures cause structural transformations, sintering and recombination of the active components, leading to a more rapid deactivation, which might be irreversible. Therefore, the development of new MDR processes is needed, including new catalytic systems that inhibit the kinetic of coke formation and promote the reaction at lower temperatures. Regarding coke formation inhibition, the use of highly dispersed metal species based on the concept of ensemble size control has been proposed. Coke formation has been shown to decrease, by downsizing metallic particles on the support. New catalyst synthesis methods have been developed to produce nanostructured precursors, which after modification give rise to homogeneously disperse nanoparticles, distributed through the structure of the support. The resulting catalysts exhibited increased activity and stability [88, 89].

Techno-economic analysis of MDR combined with other processes has been performed. Among these processes, the MDR combined with the oxidative coupling of methane (OCM) process has been analyzed. The considered data was experimentally obtained in a mini-scale OCM plant and alternatives of integrated configurations were screened. The performance of intermediate and downstream adsorption unit was investigated [90]. An economic comparison of methanol production either via SMR or MDR indicated that MDR could be a competitive process pathway providing a negligible cost of CO$_2$ import. Competitiveness resulted from lower operating and capital costs [91]. Techno-economic analysis for MDR in a membrane reactor (MR) was also conducted by using process simulation and economic analysis. Results showed economic viability of producing ultra-pure H$_2$ while simultaneously utilizing CO$_2$, by the use of MR [92].

A technology that combines CO$_2$-rich natural gas dry reforming with Fischer–Tropsch to olefins technology to produce high value-added linear α-olefins products has been developed. Processing of rich CO$_2$ natural gas causes a large amount of energy loss and carbon emission. The evaluation of the energy efficiency, carbon efficiency, and economic performance of this integrated system showed that the net CO$_2$ emission of this system is only 0.06 ton$_{CO2}$/ton$_{product}$ and the carbon efficiency of the system is up to 60.8% [93].

A recent review has identified some of the opportunities for the development of chemical processes for valorization of CO$_2$, which can be directly integrated to the energy source, in particular, for integrated nuclear-chemical energy systems [94]. The solar energy has also been explored for the integration of the dry reforming process, considered as the thermochemical storage and RE power transmission. The proposed Solchem process [95, 96] and the Closed Loop Efficiency Analysis (CLEA) Project at Sandia National Laboratories [97] represent examples of solar-chemical integrated energy systems (IES), in which the concept of the chemical heating pipe is applied (see Section Source of energy).

Among the analyzed technologies, two processes have reached industrial scale: the Sulfur PAssivated ReforminG (SPARG) process and the CALCOR process [98, 99] though their deployment has been very limited. The SPARG process is based on the selective sulfur poisoning of the most active sites [100], as found by Rostrup-Nielsen upon studying SMR Ni catalysts [101]. In fact, sulfur might act as a disassembling agent that blocks metal clusters to preclude multi-atomic interactions, inhibiting coke formation. Thus, the sulfur passivating effect operates on both reactions, reforming, and carbon formation. Pilot-scale testing of a configuration that included feedstock desulfurization, pre-reforming, and primary reforming demonstrated the carbon-free feature [102]. Later on, an industrial-scale demonstration unit was procured by revamping an SMR plant in Texas (USA) to CO_2 reforming. The SPARG technology was commercially demonstrated during a continuous 4-year run, in the Texas plant, confirming the need for a pre-reformer and the potential of a carbon-free operation [100].

From the point of view of an industrial perspective, high hydrocarbon production, reducing carbon formation, requires the use of large amounts of water as co-feed, as previously mentioned. Up to 133,000 Nm^3/h were produced using traditional nickel catalyst. Better results were attained with the SPARG process or the use of noble metal catalysts under severe conditions [103].

Among the processes that can contribute to the development of less contaminating energy supplies, the MDR to syngas by the use of biogas emerge as a possibility. Biogas is a gas generated by the anaerobic digestion of biodegradable matter such as energy crops, wastes, and residues by biodegradation reactions of organic matter. There is an increasing interest in producing biogas to reduce GHGs emissions and to achieve a sustainable development of less contaminating energy supply. Several fermentation processes are applied for biogas production, via wet and dry fermentation systems, with wet digester being the most applied one, and with different stirrer types. It depends on the origin of the feedstock [104]. The biogas composition contains mainly CH_4 and CO_2 in the order of 60–40%. Syngas production from biogas via MDR shows great potential as a green alternative to meet global environmental goals, with nickel/alumina catalysts being the most popular choice for the reaction. However, the syngas production costs estimated from a techno-economic analysis revealed syngas prices ranging from 1.15 to 1.56 €/m^3 to reach profitable setups comparatively higher than fossil fuel produced syngas (0.08 €/m^3) [105]. The effect of parameters such as pressure, temperature, and reactant composition on the reaction performance has been studied. It was observed that methane and carbon dioxide conversions can be improved and that a higher hydrogen/carbon monoxide ratio can be attained by increasing the amount of air [106]. Currently, biogas is mainly utilized in engine-based combined heat and power plants; other applications include biofuel, methanol, and electricity.

2.3.1.2 Thermochemical Cycles

Thermochemical cycles solely combine heat sources with chemical reactions to split a molecule into other components. The term cycle is used because the chemical compounds present in these processes are continuously recycled [107]. Chemical looping is a good example of a thermochemical cycle, originally conceived as a set of reactions that enabled oxygen transport within the reacting system. In chemical looping, a redox process is carried out cyclically using a solid material to circulate oxygen from one set of reactants to the regeneration step in which the original solid material is restored. Numerous chemical-looping-based process concepts have been investigated to provide feasible options for technologies development. Three concepts have been illustrated by Buelens et.al. [108] and are briefly summarized in Table 2.2: chemical—A, thermochemical—B, or carbonation—C.

Some applications of these process concepts on CCU pursue the direct conversion of CO$_2$ into CO, which is deemed an economically sustainable pathway, while others try to concentrate or recover pure CO$_2$. Chemical looping is a very promising CC method that reduces energy and cost penalties for CC. The Chemical Looping Combustion (CLC) concept is based on the transfer of oxygen from the combustion air to the fuel by means of an oxygen carrier in the form of a metal oxide, thereby avoiding the direct contact between fuel and air [109]. Chemical-looping combustion (CLC) has been gaining traction as an option for CO$_2$-free electricity generation from fossil fuels since it includes GHG separation. The CLC system comprises two interconnected reactors usually denominated as air reactor and fuel reactor (Fig. 2.4). In the fuel reactor, an oxygen carrier (usually a metallic oxide) is used to totally or partially oxidize the fuel, as depicted in reaction (2.17). In the case of total oxidation, the gas leaving the fuel reactor contains mainly CO$_2$ and H$_2$O. After water condensation, captured CO$_2$ is ready to be sent to utilization or storage. The exit gas stream from the fuel reactor contains CO$_2$ and H$_2$O. After water condensation, almost pure CO$_2$ can be obtained. Therefore, there is little energy lost in component separation. The metal generated via the reduction of the metal oxide, Me, is then transferred into the air reactor where it is regenerated by taking up oxygen from the air (Reaction 2.18).

$$CH_4(CO, H_2) + MeO \rightarrow CO_2 + H_2O(CO_2, H_2O) + Me \quad \text{Fuel reactor} \quad (2.17)$$

Table 2.2 Examples of chemical-looping concepts

		Hemi-cycle 1	Hemi-cycle 2
A	Chem.-Redox	Me + Oxidant → MeO + Reductant	MeO + Reductant → Me + Oxidant
B	Therm.-Redox	Me + Oxidant → MeO + Reductant	MeO + Heat → Me + ½O$_2$
C	Carbonation	MeO + Diluted CO$_2$ → MeCO$_3$ + Diluent	MeCO$_3$ + Heat → MeO + Pure CO$_2$

Fig. 2.4 Typical
chemical-looping process

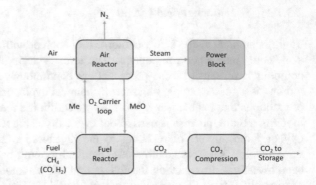

$$Me + 1/2O_2 \rightarrow MeO \quad \text{Air reactor} \quad (2.18)$$

A review article reports the advances achieved by the extensive research on oxygen carrier development, reaction kinetics, reactor design, system efficiencies, and prototype testing, using transition metal oxides. The lack or low experience on building and operating large-scale CLC systems has created barriers for the commercial implementation of this technology [110]. Chemical looping may offer a high efficiency option for reducing costs and energy demands from the power and chemical sector. In carbonation chemical-looping systems, energy efficiency is maximized at higher pressures. Thus, system design, operation, and scale-up are more complex than it is for other systems, operating at milder conditions. Although pressure has been found to negatively affect the chemical-looping kinetics of C-capturing step, it is effective on process intensification. Different reactor configurations have been examined to determine their operability and identify potential technical challenges. A lack of relevant information among the six configurations examined left doubts for a robust selection. More information is needed on the technical feasibility of using high-temperature valves and filters at system downstream, oxygen carrier longevity, and the challenges for upscaling. TEA results indicated that the reactor temperature creates constraints for applications on combined cycle power generation [111].

It is interesting to notice that the chemical-looping technology can be successfully integrated in an integrated gasification combined cycle (IGCC) plant [112]. In this case, syngas (CO + H2) is used as fuel in the looping cycle. Reactions (2.19), (2.20), and (2.21) exemplify the chemical reactions that are carried out in a syngas-based looping cycle using iron oxide as oxygen carrier.

$$Fe_2O_3 + 3CO \rightarrow 2Fe + 3CO_2 \quad \text{Fuel (syngas) reactor} \quad (2.19)$$

$$3Fe + 4H_2O \rightarrow Fe_3O_4 + 4H_2 \quad \text{Steam reactor} \quad (2.20)$$

$$4Fe_3O_4 + O_2 \rightarrow 6Fe_2O_3 \quad \text{Air reactor} \quad (2.21)$$

A chemical-looping dry reforming performed in three stages (methane reduction, CO_2 reforming, and air oxidation, see reactions 2.22, 2.23, and 2.25) has been investigated:

$$4MeO + CH_4 \rightarrow 4Me + CO_2 + 2H_2O \quad \text{Methane reduction} \quad (2.22)$$

$$2Me + 2CO_2 \rightarrow 2MeO + 2CO \quad \text{CO}_2 \text{ reforming} \quad (2.23)$$

$$2Me + 2H_2O \rightarrow 2MeO + 2H_2 \quad \text{Steam reforming} \quad (2.24)$$

$$CH_4 + CO_2 \rightarrow 2CO + 2H_2 \text{ Net reaction (R.22 + R.23 + R.24)} \quad (2.25)$$

$$2Me + O_2 \rightarrow 4MeO \text{ Air oxidation} \quad (2.26)$$

The products of this chemical loop involving the 2.22, 2.23, and 2.25 reactions are CO and H_2O, rather than syngas [113–115]. Producing syngas from this type of chemical looping would require additional incorporation of the 2.24 reaction, modifying the whole process to a bi-reforming process represented by the net reaction 2.25. A spinel nickel ferrite ($NiFe_2O_4$) required a temperature of 900 °C for the air oxidation step [113]. A study on CeO_2-modified Fe_2O_3 reported a maximum CO yield of 20 wt% [114]. The mixture of CeO_2 and Fe_2O_3 was more active than each of the individual oxides. Thermodynamic feasibility of chemical-looping dry reforming was validated through screening calculations [115]. As a whole, high CO_2 conversion is feasible and probably iron-based candidates might be the most promising.

Another promising looping system is based on calcium sorbents [112]. The calcium-based sorbent is circulating between the carbonation reactor (where CO_2 is captured shifting also the WGS reaction equilibrium to the right) and the calcination reactor (where $CaCO_3$ is decomposed for sorbent regeneration).

Once pure CO_2 is obtained, several reactions may be carried out aiming at the fixation thereof. Solar-driven decomposition of CO_2 into CO over ceria-based catalysts is one of them [116]. An open loop includes the Boudouard reaction 2.16 followed by the WGS reaction 2.10 to use part of the produced CO in rendering hydrogen [117]. This open loop would provide a flexible way of varying the H_2/CO ratio, according to the downstream processing requirements.

Also, an interesting possibility is the sorption-enhanced Boudouard (SEB) reaction, which uses CaO as both a catalyst and a CO_2 sorbent. Production of pure hydrogen can be effectively achieved through this route, which involves a highly selective removal of CO and CO_2 from a hydrogen-rich product stream [118]. In fact, based on Le Chatelier's principle, the in situ CaO carbonation removes CO_2 from the reacting media, enhancing conversion. The reaction scheme is expressed by reaction 2.27:

$$CaO + CO_2 \leftrightarrow CaCO_3 \quad \Delta H_{298K} = -178 kJ/mol \quad (2.27)$$

Table 2.3 Examples of closed-loop cycle reactions

Reaction	Temperature (K)
$2Cd + CO_2 \rightarrow 2CdO + C$	600
$2CdO \rightarrow 2Cd + O_2$	1900
$2Zn + CO_2 \rightarrow 2ZnO + C$	1300
$2ZnO \rightarrow 2Zn + O_2$	2400
$2CoO + 2SO_2 + CO_2 \rightarrow 2CoSO_4 + C$	500
$2CoSO_4 \rightarrow 2CoO + 2SO_2 + O_2$	1400

The combination of reactions 2.16 and 2.27 would give the following overall reaction 2.28:

$$CaO + 2CO \leftrightarrow CaCO_3 + C \quad \Delta H_{298K} = -350 kJ/mol \quad (2.28)$$

Finally, some chemical reactions were proposed to close the loop, by producing carbon (C) while incorporating additional CO_2 in the attempt. Direct reduction of CO_2 as well as secondary reduction via CO were considered. These reactions are collected in Table 2.3. At the time, these reactions were proposed, solar energy (the recommended source) was in its development infancy. Nevertheless, potential challenges and problems were identified, particularly regarding energy efficiency and economy of the processes.

2.3.1.3   Co-electrolysis of CO_2 and Water

The electrochemical reduction of CO_2 can be performed under various process conditions and either in the gas phase or liquid phase. Thus, two types of processes are being studied and developed, namely, low-temperature electrolysis (LTE) (T < 100 °C) and high-temperature electrolysis (HTE) (T ≤ 600 °C) processes [119]. Processes operating at moderate or intermediate temperature ranges are based on ion conductive materials, featuring adequate conductivity at the desired temperatures.

Meanwhile, gas phase processes were nurtured by the development of gas diffusion electrodes (GDE). These processes are typically operated within the LTE conditions ranges. GDEs facilitate the mass transfer of the gas to the catalytic site while the water flow can be controlled separately, thus inhibiting hydrogen evolution [120]. On one side, high CO_2 concentrations are required for avoiding GDE flooding (deactivation), and on the other, these conditions lead to low conversion [121, 122].

Relevant to the present work, co-electrolysis of CO_2 and steam to produce syngas is the most representative case of HTE [123]. Using the generated H_2 for the RWGS reaction (2.12) has also been considered [124]. CO_2 reduction at HTE conditions showed better current density standards than LTE for the production of CO, mostly because of the thermal energy contribution [119]. The use of high temperatures creates opportunities for the production of other compounds beyond CO, for the development of intensified processes and of new process configurations and combinations, e.g., electro-catalysts and heterogeneous catalysts, hydrogen generation for in situ catalytic hydrogenation of CO_2, etc. A combination of chemical and electrochemical processes, for the total reduction of CO_2 in HTEs, is a potential application of a combined type of processes [125].

Syngas could be produced via CO_2 reduction though CO production through this process is not a well-established technology. High temperatures are required for the reduction; also, this reaction presents equilibrium limitations, problems regarding carbon deposition, and the generation of undesirable hydrocarbons. Hence, here electrochemical CO_2 reduction presents an opportunity that when carried out in combination with water electrolysis aims at yielding syngas [126]. Although C1 organic molecules such as methanol can be preferably produced via catalytic process, the case of syngas seems to be an exception, giving room to alternative technologies such as electrochemical CO_2 reduction.

The lowest cost of H_2 attained via CO_2 co-electrolysis mode under HTE was of \$0.660/gge (gallon of gasoline equivalent). Compared to LTE, the energy consumption of an HTE can be 30–35% lower [127]. However, under HTE conditions, material integrity and stability are lower than under LTE conditions. Therefore, the concerns on the long-term stability of any HTE electrolysis system need to be addressed and sort out [128–131]. These concerns or probably the lack of solutions might be the reasons for the limited number of industrial HTE operations [119].

R&D activities have been trying to achieve current densities over 200 mA/cm² at high faradaic efficiencies. An example is provided by the production of CO and formate, a two electron transfer reaction [121]. The Faradaic efficiency (FE) in the CO_2 electrochemical reduction is measured as the ratio of the amount of produced CO (or any other main product) and its theoretically produced amount (from Faraday's law):

$$FE(\%) = \frac{F n_i X_i f_{CO2}}{I}$$

where n_i is the number of the electrons involved in the considered CO_2 reduction, F is the Faraday constant, X_i is the volume fraction of the considered product, I is the electrical current, and f_{CO2} is the molar CO_2 gas flow rate.

One of the cost factors negatively impacting electrocatalytic applications is the need for a pure or highly concentrated CO_2 stream. Therefore, both capture and purification processes are required to produce a suitable CO_2 feedstock. So far, studied processes at scalable electrocatalytic conditions exhibit low conversion values. Therefore, additional steps for products separation and purification, and for recovering and recycling unconverted reactants are needed and will also negatively impact the process economy [119, 132, 133]. Clearly, opportunities for process intensification and/or cost reductions are open. Intensified processes for the direct conversion of captured CO_2 are under development. Examples are an LTE co-electrolysis process [134], the production of syngas at 70% conversions and current density of

200 mA/cm^2 [135], and moderate current density (50 mA/cm^2) with FE of 72% [136].

Currently, a fast growth in the number of patents and patent applications for electrochemical and electrocatalytic reduction of CO_2 has been observed [137–144]. Such increasing number of patents is giving rise to emerging start-up companies using this growing intellectual property. In fact, recent publications compared emerging CO_2 electrochemical conversion technologies and scale-up attempts [145, 146].

2.3.2   CO₂ Hydrogenation

Hydrogenation of CO_2 not only contributes to reduce CO_2 in the atmosphere, but it also results in production of fuels and valuable chemicals [147, 148]. The catalytic hydrogenation reactions have shown a high possibility of efficiently producing high value-added products, such as alcohols, aldehydes, esters, and acids. The hydrogenation of CO_2 to alcohols or other hydrocarbon compounds is considered an important strategy for recycling the CO_2 released into the atmosphere due to combustion processes [149–151]. However, the use of CO_2 as a chemical feedstock is limited to a few industrial processes such as urea synthesis and its derivatives, salicylic acid, and carbonates. Among CO_2 hydrogenation products, methanol, and hydrocarbons are excellent fuels in internal combustion engines and are easily stored and transported [4].

2.3.2.1   Methanol

Methanol is an organic compound synthesized from either synthesis gas ($CO + H_2$) or CO_2 hydrogenation. A continuous increase in the industrial demand for methanol justifies the efforts that have been dedicated to the study of this route [152]. It is worth mentioning that nowadays the world consumes about 85 million barrels of oil a day, and about two-thirds as much natural gas equivalent. However, in the predictable future, our energy needs will have to be generated from any available alternate source. Methanol from CO_2 appears as one viable alternative to abate CO_2, being also a convenient solution for efficient energy storage on a large scale [153]. In fact, methanol is a liquid at ambient conditions and could be handled and distributed with the same type of infrastructure by which liquid fuels are distributed today. Furthermore, methanol can yield dimethyl ether (DME) [154], a much cleaner diesel fuel substitute which can also replace liquefied natural gas (LNG) and liquified petroleum gas (LPG). Methanol can be used as vehicular fuel, it can be converted into electricity in fuel cells, and it is an important raw material in the petrochemical industry. Hence, methanol is to be seriously considered as an energy carrier for a clean and sustainable energy future.

Under given conditions, the hydrogenation reaction of CO_2 produces methanol and water, as shown in reaction 2.29. It is an exothermic reaction at room temperature that competes with another more exothermic reaction, the methanation reaction, shown as reaction 2.30. Therefore, the use of very selective catalysts is required when methanol is the desired product. However, very little effort has been devoted to the search of a specialized catalyst for methanol production from CO$_2$ and most of the published work concerns conventional methanol catalysts (those producing methanol from syngas).

$$CO_2 + 3H_2 \rightarrow CH_3OH + H_2O \quad \Delta H_{298K} = -11.8 \text{kcal/mol} \quad (2.29)$$

$$CO_2 + 4H_2 \rightarrow CH_4 + 2H_2O \quad \Delta H_{298k} = -165 \text{kJ/mol} \quad (2.30)$$

Abundant published literature claims that conversion of CO$_2$ into methanol by catalytic hydrogenation is one of the most promising processes for stabilizing the atmospheric CO$_2$ level. The use of a zeolite MR, for instance, gives not only a higher CO$_2$ conversion but also higher methanol selectivity and yield when compared to results obtained with a conventional reactor [155]. Moreover, as previously mentioned, such produced methanol could be used as fuel or basic chemical to meet the large worldwide demand.

The hydrogenation of CO$_2$ to methanol is also a very interesting route to promote CO$_2$ fixation, since methanol can be transformed into olefins (e.g., ethene and propene massively used in petrochemicals and plastics production). Methanol to olefins (MTO) is the well-known MTO process [156]. Such olefins can be reacted to produce polyolefins, which are used to make many plastic materials, thereby contributing to a more long-term carbon fixation.

Finally, CO$_2$ capture and its reductive conversion into methanol which is not only an energy carrier but also a chemical platform capable of producing an enormous variety of derivatives gives rise to a new area of knowledge known as the Methanol Economy [157].

However, hydrogenation reactions as decarbonizing plan gave rise to the so-called "hydrogen conundrum" since traditional H$_2$ must not be used. The most widely used process technology for hydrogen production is the steam reforming of methane (SMR), which is both energy and carbon intense. Therefore, production of green H$_2$, via electrolysis or photolysis of water, using low- or zero-carbon energy source is mandatory [158]. Methanol synthesis licensors like Topsoe are offering their expertise on green hydrogen production and on feedstock pretreatment (for removing pollutants and poisons from the feed) for waste gas streams. Additionally, this company also offers CO$_2$ electrolysis technology, to produce CO [159]. However, the reader must be aware that an integral LCA of these process technologies is not publicly available.

Regarding electrochemical reduction into fuels, new approaches in the gas phase have been proposed. Electrocatalysts usually comprise conjugated microporous polymers (e.g., TPE-CMP) doped with Pt nanoparticles, which must act as active phase. They are mixed with carbon nanotubes (CNT) to ensure a high electronic conductivity. CO$_2$ adsorption may be strongly enhanced due to the polymer pore structure.

Such novel device provides good performances in terms of liquid product formation due to the high local concentration of CO_2 on the polymer surface where the active metal nanoparticles are deposited. Therefore, rather promising results open a new horizon for the electrocatalytic conversion of CO_2 to liquid fuels by using solar energy as well as closing the production/consumption CO_2 cycle [160]. Also, incorporation of biocatalysts to the electrochemical routes has been considered [161, 162] and the potentialities of this type of processes have been largely discussed [163].

As far as catalytic processes to produce methanol from CO_2 are concerned, two processes deserve special attention: a direct single-step and a two-step process. Regarding the single-step process, two options are commercially available: the Olah process [164] and the Lurgi process [165]. The CAMERE process is an example of the two-step process. In this process, carbon monoxide is first produced by CO_2 hydrogenation through the RWGS reaction (as shown in 2.12 reaction), over Ni-alumina catalyst [166]; then, with further addition of hydrogen, the gas stream is conveyed to a second reactor in which methanol is produced, as shown in 2.31 reaction. The inner tubes of methanol synthesis reactor are coated by a water perm-selective membrane to remove H_2O [167, 168]. The Olah technology was commissioned in 2012, following years of development. The plant is located in Svartsengi, near Grindavik, Iceland [169]. The production unit captures carbon dioxide from flue gas released by an adjacent geothermal power plant. The process implements the direct hydrogenation of CO_2 with H_2 over a mixed metal oxide catalyst, being hydrogen generated by electrolysis of water. The main reaction is shown in 2.31. Although the Olah process is currently operated in Iceland, both CAMERE and Olah processes can be considered emerging technologies. In fact, such processes (single and two step) present low conversion and poor selectivity due to the occurrence of multiple side reactions. The CAMERE process has a slightly higher yield of methanol as result of its higher conversion, in comparison to that of the Olah process. Notwithstanding, a marginal advantage in terms of economics and energy efficiency was observed for the Olah process, in contrast to the CAMERE process [170]. Economics improvement may be achieved via integration to power plants. Such integration is important since it provides a better energy use. In fact, energy management seems to be a crucial issue in these technologies. The Lurgi technology optimized energy utilization by using the heat of reaction to generate steam and export it to a power plant. The Sud Chemie C79-05-GL highly active catalyst allows the Lurgi process to optimize space/time by combining recycling unconverted feed and make-up with fresh feed.

$$CO + 2H_2 \rightleftharpoons CH_3OH \quad \Delta H_{298K} = -91kJ/mol \tag{2.31}$$

Another good example of integration is given in [171]. A power station plant burning natural gas for electricity production delivers CO_2 that is transformed into methanol. Thus, a reduction of 62% is observed in the CO_2 release to the environment. Furthermore, such integration not only recycles the CO_2 gas within methanol synthesis process, but also reduces the uptake of raw materials such as natural gas (NG). The proposed integration resulted in 25.6% reduction of the methane uptake and 21.9% decrease in the combined CO_2 emissions from the power plant and chemical plant. The reported energy sustainability indicators were: energy intensity of 33.45 GJ_{th}/ton_{MeOH} and 0.64 $GJel/ton_{MeOH}$; energy efficiency of 59% and product/feed ratio of 2.27 ton_{MeOH}/ton_{CH4} higher than 1.69 ton_{MeOH}/ton_{CH4} (calculated for a comparable standalone NG-based methanol synthesis plant with 68% efficiency) [172].

Numerous electrocatalytic hydrogenation reactions for the production of higher reduced products have been subject for patents [173–177]. The described processes of these patents concern bench-top scale exhibiting low yields and poor selectivity under high overpotential, low FE, and low current densities. A review article discusses the various tested catalysts (e.g., enzymes, carbon nanotubes, metals, metal alloys, metal oxides, and metal chalcogenides) as well as the attained achievements [163]. A bio-electrocatalytic reduction process for methanol production, using immobilized enzymes featured a FE below 10% [145, 162]. The low solubility of CO_2 in aqueous media limits the applicability of bio-electrocatalytic technologies, in high-throughput processes though a wastewater treatment application may not be highly impaired [145].

2.3.2.2 Methanation

The methanation reaction of CO_2 consists of its hydrogenation to methane. The methanation reaction, also referred to as Sabatier reaction or Sabatier process [178], produces methane and water from a reaction of carbon dioxide and hydrogen in the presence of a catalyst at high temperature, shown above as reaction 2.30. The methanation of CO_2 is an exothermic reaction, favored at relatively low temperatures and high pressures, and it is controlled by thermodynamic equilibrium. The equilibrium concentration at 1 and 30 bar for a H_2/CO_2 ratio of 4 has been reported [179], as well as the effect of pressure and temperature on conversion [180]. It was observed that high CO_2 conversions can only be achieved with a proper thermal management of the reactor, in addition to suitable selection of temperature and pressure conditions. Despite being highly exothermic, $((\Delta H)=-165°$ kJ/mol), the chemical process results intense in energy, particularly for the compression needs for achieving desirable conversion levels [180].

The greatest challenge involved in methanation is the temperature control of the exothermic reactions, meaning an efficient heat removal, which is closely linked to reactor design. In fact, one of the main issues of the process is the control of the enormous heat release which requires optimization studies [181]. The adiabatic fixed bed reactor represents the simplest reactor design option. In this case, for a stoichiometric feed mixture (20 vol% CO_2 + 80 vol % H_2) at 30 bar, the adiabatic temperature at equilibrium raises from 25 °C to 724 °C [179]. On the other hand, isothermal conditions are a prerequisite for a good catalyst performance. For the methanation of CO_2, depending on the reaction conditions, significant deviations from isothermal conditions with fluctuating conversions may be observed. Elimination of the presence of hot spot is necessary to obtain temperature profiles with a high degree of certainty [182–184]. Nickel nanoparticles (Ni-NPs) grown on electrically conductive and macroscopically shaped oxidized carbon-felt disks (OCF) heated by electromagnetic induction have been proposed [185]. Induction heating permits the electromagnetic energy to be directly absorbed by the OCF thereby converting it into heat to be transferred to the catalyst active sites, providing high conversions and selectivity to methane.

There are many other factors and reaction parameters that may influence CO_2 conversion. Methane production efficiency also depends on reactor's space velocity, reaction's stoichiometry, catalyst's surface area, and type. Regarding the catalysts used in this reaction, a very wide variety of catalytic systems is available in the literature and also in industrial processes [182, 186–191]. Nickel supported on silica or ceria-zirconia mixed oxide [188], ruthenium supported on Al_2O_3 [192] or on TiO_2 [193] have been reported as excellent catalysts capable of giving high conversions. A review of the materials and their performances together with some industrial applications may be found in [194]. These technologies are useful methods to store energy as chemical energy. Eventually, it must be borne in mind that carbon dioxide methanation is not only an alternative for chemical storage of RE but also a tool of greenhouse gas reutilization [194]. As a matter of fact, this reaction yields a product with a high energy density (methane). A crucial requirement for storage technologies is an elevated storage capacity combined with high charge/discharge periods. In that sense, only chemical secondary energy carriers such as hydrogen and carbon-based fuels (SNG) are able to fulfill this requirement [195]. As far as temperature is concerned, it has been found that the highest CO_2 conversion rates are achieved at the temperature range of 300–400 °C [195]. Also, as previously mentioned, the reactor design may have a considerable impact on the reaction conversion and operation conditions must be optimized [196, 197].

The feasibility of the Sabatier reaction has been demonstrated at laboratory scale, using renewable H_2 directly generated from water electrolysis. A temperature range of 300–350 °C optimized CO_2 conversion (60%), which occurred at a space velocity of about 10,000 h^{-1}, and a molar ratio of H_2/CO_2 of 4/1. Although increasing the H_2/CO_2 ratio improved conversion of CO_2, H_2 is most efficiently used at lower H_2/CO_2 ratios [198].

An interesting possibility concerns an innovative combination of direct air capture (DAC) with the methanation reaction in a single reactor unit through tandem steps, under H_2 flow at 300 or 350 °C [199]. In such system, a Ru/Al_2O_3 catalyst was tested together with the K_2CO_3/Al_2O_3 composite sorbent, aiming at increasing CO_2 capture from the air. The conventional Sabatier processes usually require cascade reactors and high operational pressure. However, the sorption-enhanced Sabatier reaction has the potential of producing high-grade methane at operational pressures below 10 bar. Indeed, the concept of sorption-enhanced methanation was demonstrated at atmospheric pressure with commercial Nickel-based catalyst and zeolite 4A adsorbent between 250 and 350 °C, reaching close to 100% conversion [200].

Recently, the German government has launched an environmental policy known as *Energiewende*. The *Energiewende* means the ongoing transition to a low carbon, environmentally sound, reliable, and affordable energy supply. Such new policy relies on RE, particularly wind, photovoltaics (PV), and hydroelectricity; energy efficiency;

and energy demand management. Indubitably, Energiewende is highly dependent on electric power; thus, it requires solutions to serve non-electric energy uses and to provide massive electric energy storage. Synthetic natural gas (SNG) produced with hydrogen from water electrolysis and with CO$_2$ from mainly renewable sources is a convenient approach to fulfill the requirements of this German program. Furthermore, a new reactor concept (a metallic honeycomb-like carrier-based reactor) has been developed in order to improve heat conductivity and to enable optimized operations. An SNG plant with 1 MW feed-in will be built and a fully integrated operation will be shown [201].

CO$_2$ methanation can be part of the Power-to-Gas (PtG) systems (see Section "Source of energy and energy requirement improvements", in this chapter), in which renewable or excess electric energy is used for water electrolysis to produce hydrogen [202–205]. Such H$_2$ is then combined with CO$_2$ and converted into methane (synthetic or substitute natural gas, SNG) through the Sabatier reaction using a Ru-based catalyst [202]. SNG is particularly interesting since it is a fuel with a wide proven market for power, thermal, and mobility final use applications. In addition, SNG could be stored and transported through the already existing natural gas infrastructure, thereby being a viable option for RE storage. In fact, PtG has renewed the interest for the Sabatier reaction, which due to the accessibility and low cost of natural gas never became competitive enough to be commercialized.

Moreover, biogas may also be used as feedstock. Although the CO$_2$ present in biogas could be removed, biogas could potentially be used directly as feed gas for the Sabatier reaction. In fact, very active Ni-based catalysts containing hydrotalcite-derived promoters have been used with good results [203]. Indeed, CO$_2$ coming from anaerobic fermentation may be a convenient feedstock [204]. It is important to notice that the obtained exergetic efficiency for methanation using biogas is clearly higher than that reported in literature using methanol for energy storage [205].

Another rather innovative technology is the PtG-Oxycombustion hybridization. Such technology allows the direct comparison between biogas upgrading and flue gas methanation. Upon oxycombustion, a mixture of recycled flue gas and pure oxygen is used as comburent instead of air. Hence, the large N$_2$ content is substituted by the combustion products (mainly CO$_2$ and H$_2$O), and after steam is condensed, a high CO$_2$ concentration is achieved in the flue gas. Also, the process efficiency may be enhanced if the production of oxygen via the expensive air separation unit (ASU) is replaced by the by-produced oxygen from electrolysis. By avoiding the deployment of an air separation unit, the global efficiency of this hybrid system becomes much higher [206, 207]. Finally, biomass may be selected as fuel for oxycombustion boiler to convert the process into an entirely carbon–neutral one. Thus, after methanation, the SNG produced will be equally carbon neutral. It is worth noting that utilizing volatile RE sources (e.g., solar, wind) for chemical production systems requires a deeper understanding of their dynamic operation modes. Indeed, a methanation reactor in the context of power-to-gas applications requires a dynamic optimization approach in order to identify control trajectories for a time optimal reactor start-up, thus avoiding distinct hot spot formation [208].

Undoubtedly, the future PtG process chain might play a significant role in the future energy system, mainly if biogas is used as a source of CO_2 [209]. An interesting study was published using regions of northern Germany as a case. In this study, the feasibility of CO_2 conversion from biogas plants is presented along with its integration in existing natural gas grid. The results showed that with 480 biogas plants in the region, one would be able to utilize up to 0.7TWh surplus electricity which could produce 100, 106 m^3 STP of upgraded methane per year [210].

Several studies concern the optimization of process conditions aiming at obtaining a better Wobbe index (fuel gases interchangeability indicator), CO_2 content, and calorific value [211]. Such parameters were found to be controllable by the H_2/CO_2 ratio which fed the methanation reactor. In fact, an optimal H_2/CO_2 ratio may produce a gas with high calorific value and Wobbe index. Moreover, the potential deactivation of methanation catalysts due to biogas trace ammonia concentrations was evaluated. However, results indicate that no special pretreatment for ammonia removal from the biogas fed to a methanation process is required [212].

Alternative technologies concern an innovative biogas upgrading process. Such process combines a hydrate-based biogas upgrading section with a CO_2 methanation section. Hydrates are clathrates, that is to say, crystalline compounds containing methane or ethane molecules trapped within a crystal structure of water. Hydrate formation can be used in the selective separation of CO_2 from biogas mixtures. This technology allows the overall energy efficiency to be higher than the worldwide average energy efficiency for fossil methane [213]. Nevertheless, although biogas seems a convenient raw material for Power-to-Gas (PtG) systems, it must be borne in mind that the cost of H_2 remains a significant factor in determining the cost of renewable methane. Thus, under the three examined scenarios for biogas upgrading a net present value of zero, a minimum selling price (MSP) per m^3 of renewable methane ranged from €0.76 to €1.43 (using a very favorable electricity price to produce H_2 of €0.10/kWh) [214].

In addition to biogas as an alternative raw material, a mixture CO and CO_2 resulting from methane reforming has also been studied in methanation. Such mixture requires alternative methanator designs and methanation catalyst. Based on experimental results, new catalytic systems have been proposed to operate at high-temperature, high-pressure methanation [203, 215]. Moreover, a new methanation pathway concerning plasma generation via electrical discharges on CO2 + H_2 gas mixtures has been investigated. The utilization of plasma-assisted catalyzed reactions is an important and innovative area in chemical engineering and its application in methanation is an emerging but promising technology [215].

In biological methanation, microorganisms are used to catalyze the Sabatier reaction. Biotechnological routes include a variation of the Wood-Ljungdahl pathway, using methanogenic Archaea [216] while metabolization of CH_4 itself can be carried out by methanotrophs via ribulose monophosphate pathway or serine pathways [217, 218]. This biological reaction can be carried out at lower temperatures and pressures than those using a chemical catalyst, also having a higher tolerance to contamination present in the CO_2 source. Preliminary results are interesting and encouraging [219].

Finally, the production of synthetic natural gas (SNG) can be achieved through integrated plants in which high-temperature electrolysis is performed followed by syngas methanation. Such route deploys high-temperature electrolysis with solid oxide cells (SOEC) technology. In a rather innovative manner, the system comprises co-electrolysis of water and carbon dioxide coupled with TREMP™ (Topsoe recycle energy-efficient methanation process). Interestingly, the heat recovered from the exothermic methanation section provides sufficient energy to vaporize and superheat the electrolysis water. Hence, an important thermal integration exists between SOEC and syngas upgrade catalytic section. For that reason, the co-electrolysis plant shows a high efficiency of lower heating value (LHV) [220].

The future for Sabatier conversion of CO_2 into methane seems very promising. At any rate, a techno-economic evaluation must always be provided, outlining opportunities and challenges [221]. Nevertheless, significant cost reductions for electrolysis as well as for methanation during the recent years have been observed and a further price decline until 2050 is estimated if cost projection follows the current trend [222].

2.3.3 Carboxylation Reactions

Similarly, to the inorganic valorization, the whole CO_2 moiety can be directly incorporated in organic molecules to form acids [223], esters, lactones, carbamates [224–226], and carbonates [227]. Comparatively speaking, carboxylation is a chemical reaction in which a carboxylic group (-COO) is produced or transferred by treating a substrate with carbon dioxide. Thus, direct incorporation of a -COO- moiety in other molecules (organic or inorganics) is typically less energy intense than reaction requiring reductive conversion. This latter route required as co-reactants, compounds with a high energy value, like strained cyclic molecules, unsaturated compounds, or just hydrogen. Therefore, overall energy analysis is required, and the development of standard methods and metrics needs to be pursued.

Carboxylic acids can be synthesized directly from olefins or alcohols, CO_2, and H_2, via hydrocarboxylation which consists of a combination of the RWGS reaction (2.12) and a hydroxycarbonylation step. The overall reaction has been found to be promoted by a rhodium catalyst through the described two-step mechanism [228]. In general, this two-step mechanism, involving the RWGS reaction as first step, has been proposed as the basis for new pathways of carbonylated products, as exemplified. The RWGS reaction as a CO source from the CO_2 and H_2 pair has been recognized with high synthetic potential. The reversibility of the WGS reaction (2.10) and the well-documented catalytic details open numerous opportunities for the interconversion of the CO_2/H_2 and the CO/H_2O pairs through the WGS equilibrium [229], as illustrated in Fig. 2.5.

CO_2 reacts with alcohols, under base-catalyzed conditions, for producing carbonates. The reaction requires high pressures and yield dialkylcarbonates, according to reaction 2.32. Catalysts based on Sn (IV) and Ti (IV) alkoxides [230] as well as tin oxide and zirconia [231] have been studied, for promoting this reaction.

Fig. 2.5 Examples of synthetic pathways from the CO_2/H_2 pair

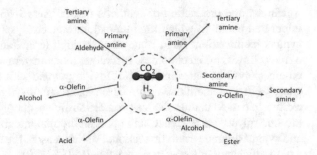

$$CO_2 + 2ROH \rightarrow RO - COO - R + H_2O \qquad (2.32)$$

The industrial demand for dimethylcarbonate has been growing significantly in recent times, underpinned by its chemical properties and its non-toxicity. The downstream applications and uses include the production of aromatic polycarbonates, its use as solvent and fuel [232]. The most common use of dimethylcarbonate is as methylating agent, which typically releases CO_2. Meanwhile, applications in polymer production would fix the CO_2 for long periods of time. Dimethylcarbonate reacts with aliphatic diols to form oligocarbonate diols, which are then used in the production of polyurethane. Dimethylcarbonate can be employed in the conversion of amines to carbamates [233], which then can be transformed to isocyanates, for polyurethane production. Further reaction of isocyanates with CO_2 to form oxadiazinetriones can expand the polyurethane chemistry [234]. Reaction 2.32 can be performed with epoxides, producing organic carbonates or polycarbonates. A great variety of catalytic materials has been tested and proved to be effective promoters of these reactions, e.g., zeolites, metal complexes, alkali metal halides, organic bases, ionic liquids, and metal oxides. None of these studied materials has demonstrated acceptable activity, stability, and/or recovery though some processes have been tested at relevant industrial scale (see Ref. [235] and references therein). In organic carbonates, CO_2 results a building block, economically and environmentally convenient. In this case, the cycloaddition of CO_2 renders cyclic carbonates. These compounds are used as aprotic polar solvents, electrolytes, building blocks of polycarbonates and other polymers, and fine chemical production. Cyclic carbonates react with bifunctional primary or secondary amines to form urethane groups that can give rise to thermosetting plastic networks [236]. There is a strong influence of the anion used in the incorporation of CO_2 in the polymer obtained during the copolymerization of CO_2 and cyclohexene oxide catalyzed by zinc. This influence reflects not only in the type of products obtained, but also affects the selectivity and kinetics of the reaction. The products obtained vary from polyethercarbonates, where short polyester sequences alternate with carbonate bonds, to polycarbonates with a strictly alternating sequence of repeating units [237, 238]. The chemical use of CO_2 as feedstock to produce valued polymers, as well as the direct polymerization of CO_2, lies in the variety of properties shown by polymers, which makes them commonly present in both simple and high-tech conditions. CO_2 can be used either directly by copolymerization or indirectly by transforming building blocks obtained from CO$_2$ in previous stages. Epoxide/CO$_2$ copolymerization to produce polyether carbonates and Bayer's industrial efforts in catalyst development and process scale-up are discussed in detail in reference [238]. Versatile materials based on polycarbonates, polyacrylates, polyurethane, etc. are emerging and applications as insulators, building materials, shoes and clothes, solar panel components, etc. have been mentioned [239]. Additional pathways for the use of CO$_2$ in the production of polymerization building blocks have been reported [223, 240].

Direct polymerization of CO$_2$ can be achieved by reaction with either epoxides to form polyalkylenecarbonates or by terpolymerization with cyclic acid anhydrides [241]. The most relevant application of polyalkylenecarbonates is long-term fixation of CO$_2$, e.g., pore-forming agents of polyalkylenecarbonates with alternating repetition units in the ceramic industry and as building materials, building foams, and insulating material of polyalkylenecarbonates with terminal alcohol groups (via polyurethanes production) [239].

Enzymes as well as biomimetic metal systems have been considered as catalysts for the synthesis of chemicals from carbon dioxide. Some promising results for various applications have been reported in the promotion of CO$_2$ (or its reduced forms) reactions, for chemicals production [242–244]. For instance, biotin enzymes catalyze carboxylation reactions, involving carbamic moieties as intermediates and also RuBisCO (ribulose 1,5-bisphosphate carboxylase-oxidase) activates CO$_2$ photosynthetically [245].

Ionizing radiation has been employed to induce carboxylation reactions since it can provoke (i) the conversion of CO$_2$ to transient species such as COO/COOH (reaction 2.33) and (ii) the formation of organic (hydrocarbonaceous) radicals, **R**· (reaction 2.34) leading to the production of carboxylic acids, esters, or organic carbonates (reaction 2.35) [246].

$$CO_2 + e_{aq}^- \rightarrow COO^- \tag{2.33}$$

$$RH + OH \rightarrow R \cdot + H_2O \tag{2.34}$$

$$COO^- + R \rightarrow RCOO^- \tag{2.35}$$

These radiation-induced carboxylation reactions have been employed to produce various organic derivatives, such as salicylic acid from phenol [247–249], malonic acid from acetic acid [249–252], and amino acids from amines [246].

2.3.4 CO₂ Incorporation in Renewable Feedstocks

The production of low-carbon energy carriers brought CO_2 utilization into considering the incorporation of renewable feedstocks. Thus, biomass, biomass intermediates, and products together with any other resource categorized as "renewable feedstock" are potential materials of interest. In this regard, municipal solid wastes (MSW) and plastic wastes in some countries fall into the renewable category.

Recently, the concept of Integrated Biorefinery [253–258] has been a debated subject and different viewpoints have been proposed [259, 260]. An efficient integrated biorefiery is the one which is capable of processing a wide range of diverse types of biomass feedstocks into value-added biofuels, renewable products, and biopower. Similar to conventional oil refineries, integrated biorefineries would render multiple products, by optimizing feedstock use and production economics. Mature technologies in refineries support residue management but in integrated biorefineries that is an existing problem which needs to be addressed.

Several types of Biomass Residues may be found. Among them, agricultural biomass, biomass from animal waste, forest biomass, and MSW deserve special attention. Regarding agricultural biomass, it could be generated from field agricultural crops (stalks, branches, leaves, straw, waste from pruning, etc.) and biomass from the by-products of the processing of agricultural products (residue from cotton ginning, olive pits, fruit pits, etc.). Furthermore, biomass by-products resulting from chemical processes using biomass are also important potential raw materials.

As previously mentioned, the chemical fixation of CO_2 may be achieved via the formation of carbonates. Organic carbonates appear as one of the main products that can be obtained from CO_2 [261]. The reaction of oxiranes (compounds containing a three-membered ring: an oxygen atom and two carbon atoms, the latter in sp3-hybridized states, e.g., ethylene and propylene oxides) with CO_2 produces cyclic organic carbonates. These reactions, however, require specific catalysts, respectively, in the presence of different type of catalysts [262]. For instance, high conversion and selectivity are obtained when polyfluoroalkylphosphonium iodides are used as catalyst in the reaction of supercritical CO_2 with propylene oxide to yield propylene carbonate [263]. In fact, the Japanese company Asahi Kazei Chemicals is currently producing polycarbonates through this route, which avoids CO_2 emissions of 1750 ton per every 10,000 ton of produced polycarbonate [264]. Application of metal-complex catalysts allows direct carbonation of methanol to render dimethyl carbonate (DMC) [265]. DMC has been used to produce polycarbonates and is an efficient octane booster additive [266].

The world is facing an increasing glycerol surplus, as a consequence of being a by-product from triglycerides transesterification used in biodiesel production [267]. This surplus leads to a continuous drop in the glycerol price which makes this molecule an attractive platform for chemical processes. Hence, glycerol needs to be transformed effectively and efficiently into value-added products. For that reason, considering glycerol as a feedstock for the production of different renewable chemicals has been the subject of many studies [268, 269].

Glycerol carbonate (GC) is a very important product in this category whose potential applications are finding space in the market. Currently, it is mostly used as solvent, emulsifier, and chemical intermediate. It is useful in producing polymers such as polyesters, polyurethanes, and polyamides which have higher market value than glycerol. Apart from that, GC is also a valuable compound for the production of glycidol which is widely utilized in pharmaceutical, cosmetic, and plastic industries [270]. GC is typically prepared using zinc salts as catalyst for the reaction of glycerol with urea [271]. Both lanthanum oxide [272] and metal-impregnated zeolites [273] have also been used as promising catalysts. As previously described in Sect. 2.2, urea is produced via the reaction of CO$_2$ and NH$_3$, being the largest CO$_2$ conversion process by volume in the industry. Thus, the use of urea in the synthesis of GC represents an indirect route to promote CO$_2$ fixation.

The direct carbonation of glycerol is possible; nonetheless, published literature is still lacking more research. The use of tin complexes as homogenous catalysts in the reaction of glycerol with CO$_2$ has been reported [274, 275]. However, at the considered conditions (50 bar, 180 °C, and 15 h), the yield of GC was rather low (5.5%). Zeolites (e.g., 13X) can be used for the preparation of GC, in supercritical CO$_2$ as demonstrated in Ref. [276], through the reaction of glycerol with ethylene carbonate. Again, results were not promising.

Ionizing radiation has been used as a facilitator for inducing reactions of CO$_2$ with renewable feedstocks [277]. The thermochemical conversion of biomass producing a gas stream bearing CO, H$_2$, and CH$_4$ is typically carried out at temperatures above 500 °C. Experiments comparing the thermochemical conversion of irradiated biomass with the untreated material showed that while irradiated biomass only produced methane, the untreated biomass yielded both CO$_2$ and CH$_4$ at different concentrations. In both cases, formation of amorphous structured carbon particles and nano-carbon black particles was observed [278]. Irradiation was also employed to induce mutations in *Chlorella sp.* microalgae that led to improvements of the biomass and/or lipid yield. Additionally, the quality of the produced lipids was improved with irradiation since the relative yields for mutated microalgae were lower for short-chain saturated fatty acids and higher for long-chain unsaturated fatty acids, compared to those produced by the untreated microalgae [279]. Irradiation enhances the reactivity of lignin, facilitating its deconstruction [280]. Similarly, formation of ·OH radicals was detected during the radiation-induced degradation of calcium lignosulfonate, which renders CO$_2$, H$_2$O, and sulfates as final products [281]. An irradiating pretreatment of cellulose could reduce its decomposition temperature producing value-added organic products [282]. As it has become apparent, the combination of low-carbon energy sources with biotechnology, catalysis, and renewable resources is a clear move through net-zero pathways, which could make significant contributions to CO$_2$ mitigation during and beyond the energy transition. New and more energy-efficient processes are needed to achieve a better exploitation of the renewable resources and to improve CO$_2$ fixation by algae.

2.3.5 Artificial Photosynthesis

The molecular aspects of natural photosynthesis converting CO_2 and H_2O to carbohydrates and oxygen have been taken as inspiration for developing chemical pathways that mimic a biochemical reaction in its different steps and use sunlight to drive it. These photochemical-based processes are categorized as "artificial photosynthesis" and have been studied for decades. In the long term, the artificial photosynthesis advances achieved through the years would emerge as a major technological breakthrough for new CO_2 conversion technologies, needing air and sunlight to produce chemicals and fuels. The intrinsic nature of these processes is hybrid, i.e., the combination of physical (photo-), chemical (electro- or catalytic), and/or biological (enzymatic, microbial, or algae), in the form of bio-electrocatalytic, photobioelectrochemical, photochemical, photocatalytic, photobiocatalytic, photoelectrochemical, photoelectrocatalytic, and photobioelectrocatalytic processes. The interested reader is directed to more specialized review articles and book chapters that have been published recently [283–295].

Bioelectrochemical synthesis is a technique that uses electro-autotrophic microorganisms under a constant applied potential to reduce CO_2 to multicarbon products, which can become a potential approach for the production of high value-added products from both organic and inorganic carbon. *Citrobacter amalonaticus*, a facultative anaerobic bacterium, has been applied as a biocatalyst in a bioelectrochemical system to produce succinic acid, a C4 molecule employed as precursor for several industrially important products in the pharmaceutical industry, agricultural sectors, food industry, and the production of biodegradable materials and polyesters. The influence of variables such as applied potential, bioelectrochemical system, and amount of CO_2 was evaluated; the highest succinic acid production (14.4 g/L) was observed at (-0.8 V) an applied potential, with the absence of CO_2 resulting in formation of higher acetic acid (6.34 g/L) and lactic acid (6.13 g/L), revealing a direct dependency of succinic acid production on the availability of CO_2. The use of carbonate as a CO_2 source was also investigated, and a positive effect on the succinic acid production (13.6 g/L) was observed. The results suggest that the biological route is capable not only of producing industrially valuable chemicals but also fixing CO_2, making the process environmentally friendly [296].

In a recent review, photochemical, photoelectrochemical, plasma, and microbial electrosynthetic options have been presented and discussed along with proposals for future research needed from an industrial perspective [297]. This review presents various technological approaches related to carbon dioxide conversion, their current status, and the molecules for which each approach is most suitable for CO_2 reduction by means of such processes.

The potential of artificial photosynthetic processes for driving current linear supply chains towards a circular economy, using as main ingredients water and CO_2, could underpin the basis for sustainable chemical and fuel industries, mitigating the effects of climate change for future generations. Unfortunately, the efforts on artificial photosynthesis have been fading away and the idea of an economy based on

CO_2 and water appears remote, almost a dream. Nevertheless, there is always place for science and technology to realize it, but resources and reinvestments are needed immediately. The advances in solar energy and concentrated power reactors (e.g., Refs. [116, 298]) are closing the knowledge gaps, and opening new opportunities for the future. These new additional investments and efforts in the area could be seen as the basis for a sustainable future for humankind and for the planet.

2.4 End-Products and End-Users

Industries with profiles characterized by being energy intense, C-intense, or both, such as steel, cement, chemicals, oil and gas, health, etc. are directly involved in the definition, planning, or implementation of decarbonization strategies. These industries employed the most energy-intensive processes, accounting for around 60% of industrial energy demand [299] and an even higher share of direct CO_2 emissions [300]. One common factor characteristic of the industrial sector is probably the typical use of heat in the vast majority of its processes. Another commonality for the bulk use of energy in industry is steam generation [301]. Typically, heat is provided by the CO_2-emitting combustion of fossil fuels though selected fuel varies among the industries. For instance, primary steel production is dominated by blast furnaces mainly fed by coal and coke, in the refining and chemical industry oil and gas are the fuels of choice, while in cement production all fossil fuels are used, with the exception of the European industry that uses wastes [15]. By 2019, the industry contributed to 30% of the global CO_2 emissions (~11 Gton) [302], associated with an energy demand of more than 175 EJ. CC plays a vital role whether the carbon gets sequestered or converted into value-added products. The emitted CO_2 industrialization passes for its capture and for conversion processes recovery in a concentrated stream since CO_2 concentration from the vast majority of large emission sources is lower than 15%, only a few (<2%) of the fossil-fuel-derived industrial sources exceed a 95% of CO_2 concentration [15].

For these concerned industrial sectors, activities for the development of new CO_2 utilization technologies are crucial. They are involved either in the development of new conversion processes or the implementation of new applications. Thus, the development of better and more efficient processes with reduced overall carbon footprints to produce added-value products are required and for which incorporation of CC, CS, and CU process technologies is a key element, for an intrinsic and synergistic connection that set the economic viability of any project. Although sometimes seen as competitors, these technologies are in fact complementary and/or supplementary and should be orchestrated to provide the best benefit. CC technologies are key in providing large volumes of high-purity CO_2, creating new opportunities for recycling and utilization. The almost unlimited availability of carbon dioxide makes its chemical conversion, uses, and applications of potential interest. However, considering the difference in scale of emitted CO_2 and that of the market of chemicals and C-bearing products, multiple new processes and new products are needed.

However, whatever the new process or new product is, a low-C or RE source should be employed. Currently, commercially available technologies are only a few and the emerging technologies are not so close to industrial realization [227].

The UK government has established that by 2050, an 80% CO_2 emissions reduction must be attained, and has conceived an approach consisting of a decrease in energy intensity of the industrial processes. Thus, the "Clean Growth Strategy", released in October 2017, mandates the decarbonization of the electricity supply and measures for an increased energy efficiency and significant energy savings through interventions into the "Circular economy" [303]. A work has identified decarbonization options for highly emitting industries (e.g., cement, steel, ethylene, and ammonia) whose common factors are deriving their emissions from feedstocks and from high-temperature heat. The suggested options were demand-side measures, energy efficiency improvements, electrification of heat, using green hydrogen as feedstock or fuel, using biomass as feedstock or fuel, CCS. The optimization of the combination of options was limited not only by economic and technical factors but also by local factors such as access to low-cost zero-carbon electricity and/or to a suitable kind of sustainably produced biomass, and availability of carbon storage capacity, for instance [302]. One of these combined strategies is adding improvements in energy efficiency of the industrial processes to the decarbonized energy supply. However, some metrics and accountability need to be implemented. The EU Directive has suggested a policy for an energy audit system, to evaluate the progress on energy efficiency. This audit system includes five steps: (1) defining key issues, (2) setting the objectives for each key issue, (3) identifying the options for each key issue, (4) analyzing options from an energy and environmental perspective, and (5) comparing options and selecting the recommended one. A proposed evaluation methodology is considered an important step towards the selection of a harmonized policy [304]. The definition of energy performance indicators (EnPIs) is essential for a successful industrial energy management, which in turn is key for improving energy efficiency. The variability of industrial processes demands the development of tailor-designed EnPIs. The pulp and paper industry was taken as an example of energy-intense industry for the development of a model and the EnPIs validation, under the assumption that such model could be generalized to other industrial sectors [305].

Mineralization via carbonation emerged as a value-adding alternative to geological sequestration, which would replace commercially driven products, by decarbonized substitutes. Carbonation represents an economically feasible mechanism for storing carbon in products and materials. As mentioned above, this technology requires further development for large-scale applications. Wet carbonation using olivine is probably the best studied case, deriving in CCS costs in the range of 50 to 100 US\$/ton$_{stored-CO2}$, due to an energy penalty on the original power plant of about 30–50%. Since the capture plant incorporates an additional 10–40% energy penalty, the whole carbonation plant incurs in a 60–180% more energy penalty than a power plant with equivalent output without CCS [15].

An illustration of the challenges in the decarbonized energy supply has been provided by an evaluation study of the Chinese power sector. The decarbonization model considered the technological progress and the cross-regional power transmission in China's power sector. The objective was to build a potential decarbonization pathway, under optimal cost constraints. The model operates at different carbon prices, in three economic growth scenarios. Power generation is undergoing different levels of change in gears depending on geographical and economic circumstances. In China, the shift goes from coal-fired plants to wind, nuclear, and hydropower. A carbon price of $21/ton was found to be optimal in promoting decarbonization, speeding up the process and reducing peak height. Once economy growth is achieved, the final cost of energy lowers and the carbon emissions reduce considerably [306].

Decarbonization of the industrial and of the energy sectors goes through intrinsically correlated activities led prominently by the reduction of fossil-fuel emissions. The sensitivity analysis performed by Kriegler et al. [307] indicated that four central options should be pursued: (i) lowering energy demand (requires more energy-efficient processes, for instance), electrifying energy services (replacing fossil-fuel-driven services, by electric services), decarbonizing the power sector (low-C or net-zero-C energy), and decarbonizing non-electric fuel use in energy end-use sectors (global decarbonization).

2.4.1 Steel and Metallurgical Sector

Steel together with the cement industry contributes to 14% of the global CO_2 emissions, which represent 47% of the total industrial sector emissions [308].

The iron and steel industry already realized what is the whole truth for global decarbonization, i.e., a single approach is not a solution. The solution varies according to the existing infrastructure, the geographical location, and the economic scenario. Current practices and available technologies could provide deep decarbonization, by leading however to significant increase in the production costs. The industry is betting for the definition of incentivizing policies that avoid the threat on job losses and emission growth [309].

Nine decarbonization approaches have been identified to reduce the steel industry emissions: (i) increasing material efficiency (better properties, longer life-cycle, and allow recycling); (ii) increasing on resource and energy efficiency (by-products/wastes and energy—heat exchanges); (iii) recycling of steel product scrap after consumer ends its lifecycle; (iv) innovation to increase sustainable value added; (v) scrap-based secondary steelmaking; (vi) development of low-emission production processes; (vii) immediate adoption of available low-C technologies; (viii) market creation for decoupling from low-C products; and (ix) immediate implementation of CCS technologies. Additionally, the important role of the business models and the needs for a deep redefinition were emphasized and generalized within industrial decarbonization [310].

The fermentation of CO_2 [311–313] has been one of the focal topics for the steel industry. New process technologies have been developed and some of them are commercially available for the biochemical conversion of the CO_2-rich flue gas streams [314–318], from steel plants. Chemolithoautotrophic microorganisms are the preferred microbial catalysts employed in gas fermentation [319]. The first autotrophic CO_2 fixation pathway was based on the metabolic pathway of the key enzyme Ribulose-1,5-bisphosphate carboxylase/oxygenase (RuBisCo) [320]. The reductive acetyl-CoA pathway found in anaerobic acetogens is considered the most efficient CO/CO_2 fixation pathway [321, 322]. Details on these metabolic pathways can be found in Refs. [323, 324].

In gas fermentation, since energy limits the acetogen metabolism, the addition of H_2 mitigates the CO_2 losses/production and improves the yield of reduced and energy-intensive products. The presence of H_2 increased ethanol yield from 15 to 61%, by reducing the feedstock losses as CO_2 from 61 to 17% [325].

Carbonation has been proposed as a mechanism to reduce C-intensity of the steel industry. Various process variables were considered for the use of industrial alkaline-rich waste (IAW) for process development of a wet carbonation of steel slag. IAW is economically available and logistically convenient to the sources of carbon dioxide emissions. A maximum of 44% of carbonation was achieved using steel slag with particle size <90 μm, at a temperature of 100 °C, CO_2 pressure of 10 kg/cm^2, liquid-to-solid ratio of 6, and reaction time of 2 h [326].

The decarbonization of the copper production industry has been hindered by its energy intensity, for which electrification falls within the economically prohibited range. An example was given on the use of green H_2, produced from water electrolysis. The evaluated cost for direct CO_2 abatement was €201/ton$_{CO2\text{-eq}}$. This abatement cost significantly exceeds European emission certificate prices, however valorization of the co-produced green-O_2 might contribute to reduce the abatement costs up to about €54/ton$_{CO2\text{-eq}}$. Finally, this study also indicated that GHG emissions would only be reduced if electricity could be generated with emission factors below 160 g$_{CO2\text{-eq}}$/kW-h [327].

The potential of mineralization for CO_2 fixation has been evaluated considering three different technologies: (i) production of precipitated calcium carbonate from steelmaking slags; (ii) blast furnace top gas fixation into magnesium carbonate; and (iii). production of metallic nanoparticles using a dry, high-voltage arc discharge process. The environmental footprint was evaluated considering emissions, material use, and reuse. Although technical feasibility was confirmed, no economic data was reported [53]. Ammonium salts can extract up to 50% of the calcium contained in the iron- and steelmaking slag (2.36), which can be used for the mineralization of the emitted CO_2 (2.37 and 2.38). The slags from the steel converter are treated through a two-stage aqueous process, to produce precipitated calcium carbonate, at ambient conditions. The performance of the process, the possibility of continuous operation, and the progress towards larger scale operation were reported and evaluated by lifecycle assessment [328].

$$CaO + NH_4X + H_2O \rightarrow CaX_2 + NH_4OH \quad (2.36)$$

$$NH_4OH + CO_2 \rightarrow (NH_4)_2CO_3 + H_2O \quad (2.37)$$

$$(NH_4)_2CO_3 + CaX_2 \rightarrow CaCO_3 + 2NH_4X \qquad (2.38)$$

2.4.2 Cement Industry

A cement is a binder, that is to say, a substance that sets, hardens, and adheres to other materials to bind them together. Cement mixed with other fine particles produces mortar for masonry or with sand and gravel produces concrete. Concrete is the most widely used material in the world. Cement may be divided into two main types: non-hydraulic cement which does not set in wet conditions or under water and hydraulic cements (also known as Portland cement) which set and become adhesive when the dry ingredients undergo reaction with water. Portland cement is by far the most common type of cement in general use around the world.

Portland cement is produced via calcination of limestone (calcium carbonate) with other materials (such as clay) at high temperatures in a kiln. Such calcination liberates a molecule of carbon dioxide from the calcium carbonate to form calcium oxide, or quicklime, which then chemically combines with the other materials in the mix to form calcium silicates and other cementitious compounds. The resulting hard substance, called "clinker", is then ground with a small amount of gypsum into a powder to make ordinary Portland cement, the most commonly used type of cement (often referred to as OPC).

The high energy intensity involved in cement production accounts for approximately 7% of global anthropogenic greenhouse gas (GHG) emissions in CO_2 equivalents [329]. Nearly one-third of the total CO_2 emissions from cement production results from the combustion of fuels, whereas the rest represents process-related CO_2 emissions resulting from the aforementioned decarbonation of limestone during calcination process (see C-hemicycle 2, in Table 2.2).

The rising global population along with infrastructure development needs to augment the demand for cement and concrete. As a matter of fact, the cement production is expected to increase 12% by 2050 [329]. Hence, there is an increasing pressure on the cement industry to reduce the carbon footprint of the production thereof.

Recently, an innovative technology entitled mineral carbon capture and utilization (MCCU) [330–335] has been proposed as a potential technology to reduce CO_2 emissions, particularly from the cement sector. MCCU is an MC technology in which calcium in alkali wastes undergoes reaction with CO_2 to form $CaCO_3$. MCCU is therefore a process where CO_2 is sequestered chemically using industrial alkali wastes such as concrete sludges and waste concretes. The MCC technology is often named "recarbonation", the product thereof being called a "recarbonate". Figure 2.6 depicts the main steps of cement production including the MCC process. Various alkali wastes have been used to capture and utilize CO_2 via carbonation. Among them, steel slag [336–338], cement and concrete wastes [339–341], and fly ash [342, 343] deserve special attention.

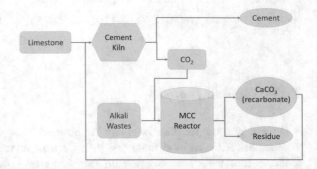
Fig. 2.6 Main steps of cement production with the MCC technology

Recarbonates may have several uses, provided some precautions are taken to prevent CO_2 emissions (leakage) from the recarbonates. In this sense, neither temperatures should exceed 825 °C nor pH should be lower than 7. The following recarbonate uses may be highlighted [344]: storage within the plant limits or at a limestone quarry; addition to cement as a minor constituent; calcination in a cement kiln as feedstock for cement production; utilization as a raw material for other productions in the plant; and sales to a third party.

Five areas were identified as critical for the cement industry to address problems and needs, in trying to reduce their CO_2 emissions [329]:

- Energy efficiency improvements.
- Fuels replacement.
- Lowering clinker/cement ratio.
- Using emerging and innovative technologies.
- Alternative binding materials for cements.

Aside from the MCCU technology, Izumi et al. have summarized a number of other measures that the cement industry may take directly, aiming at reducing CO_2 emissions [344]. These potential measures in the energy context include improving thermal energy efficiency, deploying of low-carbon alternative fuels including biomass and raw materials, of mineralizers aiming at reducing kiln burning temperature, of low-carbon binding materials, and of carbon capture and storage (CCS) technology (to be developed). Meanwhile, regarding processes their suggestions include increasing use of clinker substitutes, replacing limestone by decarbonated raw materials, applying MCCU technology using alkali wastes and applying CCS technology once developed.

As previously mentioned, MC seems to be a convenient tool for mitigating global warming through CO_2 sequestration. Indeed, gypsum, which is an industrial solid waste, is an interesting material to fix CO_2 due to its fast carbonation rate and high carbonation reactivity (above 95%). It must be called to mind that gypsum stacks are also an environmental problem. Therefore, gypsum carbonation is an interesting chemical route to mitigate both gypsum stacking and the concentration of CO_2 in the atmosphere [31].

Decarbonization of cement industry may be associated with an increase in energy efficiency. An optimization study applied to the Swiss cement industry revealed that, by combining these two strategies, future cement production will be able to reduce its specific energy consumption from 3.0 GJ/ton$_{cement}$ in 2015 to 2.3 GJ/ton$_{cement}$ in 2050. Concomitantly, cement production will decrease its CO_2 emissions from 579 kg$_{CO2}$/ton$_{cement}$ in 2015 to 466 kg$_{CO2}$/ton$_{cement}$ in 2050, by improving energy efficiency and reducing clinker content. However, economic competitiveness of capture technologies requires implementation of a minimum CO_2 tax of €70/ton$_{CO2}$ [345].

Furthermore, it is worth noticing that along with Switzerland, other countries are committed to reducing their carbon emissions. South Africa, for instance, is increasing its investments in infrastructure, which requires cement. Notwithstanding, the greenhouse gas (GHG) sectoral target for the cement industry is a reduction of 34% by 2020 below the 1990 level. To meet these targets, South Africa aims at implementing appropriate energy conservation technologies and to adopting further innovative technologies, such as the use of alternative resources, as well as Carbon Capture and Storage (CCS) [330].

Finally, one must highlight the fact that decarbonization measures with significant potential exist along the entire cementitious material cycle. However, it seems that in upstream (of use), energy and emission efficiency measures are better quantified than in downstream (of use) and material efficiency measures. As a matter of fact, the decarbonization potentials of recycling technologies and the ways by means of which technological advancements may transform the cementitious material cycle (including stocks, flows, processing of materials, and the decarbonization measure effectiveness) still require more understanding and research [346].

2.4.3 Oil and Gas Industry/Fossil Fuels

Decarbonization imposes various threats to the oil and gas industry since its CO_2 emissions are associated to its operations, products, and products uses and applications. The use of fossil fuels in the transportation sector and the application of fossil resources in energy generation are collectively responsible for more than 80% of the global emissions.

As the world gets serious about the 2050 climate neutrality target, technology deployment is speeding up. Among the emission-reducing measures available today, the International Energy Agency (IEA) expects carbon capture, utilization, and storage (CCUS) to play critical roles. CCUS is depicted in Fig. 2.7. The technologies concerning the capture of carbon dioxide (CO_2) are somewhat old (50 years). However, the concept of CCUS is newer. Such concept proposes the use of CO_2 as a raw material for several chemical transformations, which will add value to this molecule. This seems to be an efficient solution for the area of Oil and Gas. Nevertheless, opposition to these technologies is now emerging, mainly within environmental groups, which consider CCUS as an instrument for the hydrocarbon industry to remain in business. Hence, CCUS technologies in the Oil and Gas industry are a

Fig. 2.7 The concept of carbon capture, utilization, and storage

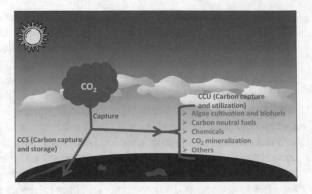

politically sensitive topic. At any rate, several actions are being carried out in many countries, aiming at implementing CCUS in their respective Oil and Gas industries. Regarding Canada, as one of the most energy-intensive processes, the Canadian in situ bitumen extraction is a huge GHG producer. Several design strategies are being considered to mitigate GHG emissions [329].

The GHG emissions generated by in situ extraction plants (e.g., steam-assisted gravity drainage facilities) can be reduced, in a large degree, by using available technologies such as heat integration and post-combustion carbon capture (PCC). Nevertheless, capital cost, energy requirements, and associated GHG emissions represent the most important current challenge limiting the CC use, in SAGD applications. However, heat recovery can provide 13–47% of the regeneration energy when the PCC technology and heat integration are considered for certain configurations of the SAGD process. Also, a capture cost of US\$34.6–55.3/ton$_{CO2}$ for different studied scenarios may be estimated [347].

China is also taking important measures to reduce CO_2 emissions. In fact, permanent sequestration of anthropogenic CO_2 in offshore sediments is a priority to achieve China's national goal of net-zero by 2060. The northern South China Sea bordering the Guangdong, Guangxi, and Hainan Provinces host the major oil and gas reservoirs, for which the CO_2 sequestration potential has been evaluated. The voidage replacement for individual oil and gas fields was the basis for calculation of the CO_2 storage potential. These calculations indicate a storage potential of 84 and 16%, for gas and for oil reservoirs, respectively [348]. China is also implementing biomass energy with CO_2 capture and storage (BECCS). As a matter of fact, co-firing biomass and coal in power plants with CCS seems to be an efficient measure for deep decarbonization in the energy sector [349]. Unfortunately, power plants based on biomass energy with CCS resulted not economically viable, unless incentive measures are incorporated.

Although several technologies for CCUS have been proposed, there is no consensus on which is the most recommendable technology. In fact, there is no U.S. EPA endorsement of any particular control strategy [350]. Furthermore, it should be noted that a number of petroleum refining processes spread across the refinery emit CO_2. It is undisputable that CCUS technologies offer significant strategic value in the transition to net zero. In fact, cement, iron, and steel or chemicals rely on CCUS to deal with their emissions, which could be used to produce long-distance transportation fuels (notably aviation). It must be highlighted that as far as the Oil and Gas industry is concerned, CCUS is not a universal panacea. Indeed, a rapid reduction in GHGs emissions, necessary to achieve the net-zero goal, can only be undertaken through a radical cultural transformation that involves the way we produce and consume energy.

Nonetheless, according to the IEA, CCUS is an important technological alternative for reducing CO_2 emissions in carbon-intensive industrial processes and offers one of the lowest cost ways of doing so [351]. In fact, Shell announces two technologies that promise cost-effectively CO_2 capturing in a wide range of applications. They are Shell's CANSOLV CO_2 Capture System (for capturing CO_2 from low-pressure streams, including flue gas) and Shell's ADIP ULTRA technology (for capturing CO_2 from high-pressure process streams).

The world economy has suffered a negative impact due to the COVID-19 epidemic. Along with this, the drive towards net-zero carbon dioxide emissions implies a total restructuration in the field of transport fuels. Alternative fuels that could be produced from CO_2, using the concept of modified Fischer–Tropsch synthesis [352], represent drop-in or replacement fuel options, for the transport sector to alleviate its dependence on fossil-carbon-based fuels. The limited availability of low-cost/high-purity CO_2 and of green hydrogen as well as catalysts with appropriate selectivity for particular chain lengths hydrocarbons are the main barriers obstructing a transition to alternative fuels [353].

Regarding synthetic fuels, economic issues must not be disregarded. The existence of an established market for synthetic fuels is undeniable. In principle, the demands of such market could be met by synthetic functional fuels, which regard their particular energy value. Thus, cost becomes the main difference between the two types of fuel since fossil fuels exhibit a considerably greater energy returned on the energy invested (EROEI) than that for synthetic fuels. Efficiency losses undergone during production of synthetic fuels are mainly responsible for their low EROEI. Instead of a low energy cost, the use of synthetic fuels ought to have additional benefits for specific market segments. Hall et al. suggested benefits on energy autonomy, and on providing an energy pathway for low-cost or off-peak electricity [354].

Nevertheless, the energy transition in the transport sector is an undisputable fact. This requires a combination of energy efficiency and energy conservation. On the other hand, it is not easy to change an entire system that has been relying on hydro-carbons for more than 100 years. The introduction of electric vehicles (EV) might be a partial solution. However, the introduction of EVs in the market represents a huge increase in the world electricity demand, which may be around 10% of the total electricity consumption forecast in 2050 [355]. Still for a low-carbon future, EVs are considered a promising technological solution. However, it is necessary to model the market penetration of EVs and determine the magnitude of the impact of EV adoption in future urban decarbonization scenarios. Factors such as location choice, land-use, transport patterns, energy profiles, and economy must be considered when implementing a stringent EV policy. Significant positive effects are expected from an ambitious market diffusion of EVs as shown by scenario simulations [356]. Whereas emission reductions will be observed in city centers, economic benefits tend to occur in suburban areas. Clearly then, the EV adoption impact would depend on the spatial organization and structure of cities.

As far as terrestrial transportation vehicles are concerned, the use of biofuels seems to be a good alternative solution for net-zero carbon dioxide emissions. On the other hand, the case of both maritime and aviation sectors is much more complex.

Fuels for aviation must meet several requirements. Considering the need for fuel to be lightweight yet having sufficiently high energy density to power long-distance flights, the most viable option today for renewable aviation fuel is that of carbon–neutral biofuels. Biofuel technology has become well established and can be blended and used in existing jet engines without modifications.

Conversely, maritime fuels must comply with different requirements and characteristics. Shipping, which accounts for 2.6% of global carbon dioxide emissions, must urgently find clean energy solutions to decarbonize the industry and achieve the International Maritime Organization (IMO)'s greenhouse gas (GHG) emission targets by 2050. The current leading options for clean maritime fuel are ammonia, hydrogen, biofuels, and methanol. However, most of these options are not yet cost-efficient and still present storage or safety issues for ship designers. Moreover, electrification of long-haul vessels is not yet commercially viable, as battery technology has not yet developed enough capacity to carry a 200,000-ton vessel across an ocean without having to recharge. Altogether, developments have yet to offer a complete solution to power large vessels across long distances while being economical, transportable, and storable at the same time [357, 358].

Ammonia is a proposed option for the shipping industry, offering relatively low GHG emissions, ubiquitous infrastructure for manufacturing, storage, and distribution, and exhibiting high energy density and competitive cost. The use of renewable energy and feedstocks in ammonia manufacture enhances its environmental benefits. Indeed, benefits are even higher when renewable ammonia is used as a transportation fuel [359]. Nonetheless, in order to provide perspectives and define a roadmap for future applications, the discussion should include topics such as environmental and safety concerns, challenges on technology adoption and adaptation, and associated costs, together with fuel switching issues [359].

In fact, new air quality regulations that legally bind the international shipping industry generated deep challenges for them. Along with ammonia, another alternative to liquid fossil fuels is LNG; however, methane emissions reduce its overall climate benefit. An interesting study [360] presents a comprehensive environmental lifecycle and cost assessment of LNG as a shipping fuel, compared to heavy fuel oil (HFO), marine diesel oil (MDO), methanol, and potential renewable fuels (hydrogen, ammonia, biogas, and biomethanol). Compared to all these considered liquid fossil fuels, LNG gives improved air quality impacts and reduced fuel costs though the benefit varies along the different LNG engine types. Nevertheless, unacceptably high methane leaks from some engines were observed. In fact, total methane emissions must be reduced to 0.8–1.6% to ensure a climate benefit over HFO. At any rate, the use of an alternative fuel is not sufficient to meet the target. Energy efficiency must increase 35% to meet a 50% decarbonization target [360].

Hydrogen is also an alternative fuel for maritime transportation. However, issues related with hydrogen storage and distribution currently inhibit its implementation [357].

The case of inland ship power systems also requires a convenient strategy. Inland shipping concerns a transport system allowing ships and barges to use inland waterways (such as canals, rivers, and lakes). This type of transportation is important for large countries with navigable waterways, such as China. The case of China has been studied and Yangtze River ships using battery power and hybrid power have been assessed. The results indicated that battery and hybrid power had lower lifetime CO_2 emissions and costs in comparison to diesel power [361].

Finally, targets for the maritime sector are undoubtedly challenging. Indeed, the 2050 targets of IMO will only be achieved via radical technology shift together with the aid of social pressure; financial incentives; and regulatory and legislative reforms at the local, regional, and international level [362].

The aviation industry must take urgent actions aiming at accelerating the transition to renewable jet fuels (RJF). Sustainable aviation fuel (SAF) can be produced from a variety of feedstocks, with 80% reduction in carbon intensity over its lifecycle. SAF is an aviation blendstock that needs to be combined with conventional fossil-based kerosene, to meet jet fuel specifications. It can be certified as Jet-A1 fuel and be used without any technical modifications to aircrafts [363].

Despite the technological breakthroughs, RJF volumes have not grown, supposedly because RJF is currently priced higher than conventional fossil jet fuel. So, a question comes up: what lies behind the higher price of RJF? RJF is more commonly known as SAF and the reasons for its higher price must be identified and addressed. Is there a lack of government support for SAF? Renewable fuels should obtain similar subsidies to those received by other renewable industries, such as electric cars and solar energy. However, this is not the case so far. Therefore, incentives are needed to foster the use of SAF [364].

Another important topic that deserves special attention regards pathways and new technologies for CCU and maximizing recovery from energy facilities as potential contributions of using biotechnology to reduce emissions. In fact, Goal 12 (ensure sustainable consumption and production patterns) of the UN Sustainable Development can be addressed through applications of microbiology in the development of petroleum biotechnologies in relation to microbially mediated energy and material recovery from unconventional sources and methods (heavy oil to methane, shale gas, and fracking), bioelectrochemical systems for electricity generation from fossil fuels, and innovations in synthetic biology, as recommended by Sherry et al. [365].

Finally, yet importantly, the fact that besides decarbonization of its operations and uses, the oil and gas sector faces the threat of the replacement of its products and/or their derived uses or applications should be considered. Indeed, future power systems with large deployment of RE sources will require broad storage capacity to ensure grid stability. In this context, power-to-chemicals is an important concept for a medium/long-term storage horizon and a wide range of capacities. Within this alternative, in a scenario whose main objective is decarbonization, ammonia might appear as a promising fuel. The first step is the production of ammonia using RE sources, followed by its transformation into energy. This thermochemical route has been evaluated, including various gas clean-up processes to valorize recoverable components and meet environmental regulations. The use of ammonia as an energy storage alternative was found to be feasible, providing a powerful platform for the deployment of power grids of these fluctuating sources type, with high penetration potential [366].

2.4.4 Health Sector

CO_2 emissions from health care in the world's largest economies account for about 5% of their national carbon footprints [367]. In fact, the combined emissions from hospitals, health services, and the medical supply chain across the OECD group of market-based economies, as well as China and India, make up around 4% of the global total. Surprisingly, this is a larger share than that of either aviation or shipping sectors.

The connections between climate change and human health are rather complex, involving three main areas. First, climate change impacts such as heat waves, storms, floods, droughts, fires, altered infectious disease patterns, air pollution, and food shortages will increase the demand for healthcare services. Second, climate-health co-benefits combine the long-term benefits of reduced greenhouse gas (GHG) emissions with more tangible and short-term benefits for public health. Finally, the health-care sector is a large and socio-economically important sector and is itself a significant cause of CO_2 emissions.

Health care is a major component of the global economy. Actually, OECD nations spend an average of 9% of their GDPs on it. Furthermore, it is an energy-intensive sector. Unfortunately, despite efforts by the Lancet Commission on Health and Climate Change [368] to promote the tracking of healthcare emissions, the subject has received little attention from the wider research community.

However, a careful analysis of the sector may not only highlight health care itself as a significant cause of emissions but may also reveal the potential to make major improvements. Such improvements will cut emissions, being also able to improve public health. In fact, climate change and healthcare sector are intimately linked, since the rising of the global temperature is associated with the spread of infectious diseases. Also, dangerous weather events may have a considerable negative impact on the world's population health.

2.4.4.1 Medical Carbon Dioxide and Carbogen

Carbon dioxide is widely used as an insufflation gas and cooling agent. In health care, mainly in hospitals, a gas called Carbogen is often deployed. Carbogen is a mixture of carbon dioxide and oxygen, being used to stimulate breathing in the treatment of respiratory diseases. Carbogen is also called Meduna's Mixture after its inventor Ladislas Meduna. Meduna's original formula was 30% CO_2 and 70% oxygen; however, the term carbogen can refer to any mixture of these two gases, from 1.5% [369] to 50% CO_2 [370].

Carbogen is used in several medical applications. It can be administrated by mask or endotracheal tube. As previously mentioned, it is used to stimulate respiration in situations such as the investigation and assessment of chronic respiratory disease, to stimulate breathing after a period of apnoea, and in the management of chronic respiratory obstruction after the obstruction has been relieved. Carbogen is also added to several anaesthetic and oxygenation mixtures used under special conditions such as cardio-pulmonary bypass surgery and the management of renal dialysis.

Hence, medical carbon dioxide (CO_2) has various medical applications. It may be used as a pure gas or in specialized mixtures with other gases. Besides the afore-mentioned use in stimulating breathing, CO_2 may also be employed in anaesthesia [371] and sterilization of equipment [372].

Furthermore, medical carbon dioxide can be used as an insufflation gas for minimal invasive surgery, such as laparoscopy, endoscopy, and arthroscopy [373]. The gas enlarges and stabilizes body cavities, thereby allowing better visibility of the surgical field. Moreover, in its liquid phase, medical carbon dioxide can be used to reduce temperatures down to $-76\,°C$, for cryotherapy [374] or for local analgesia [375]. Other applications of medical carbon dioxide include transient respiratory stimulation [376] and encouragement of deep breathing and coughing to prevent or treat atelectasis.

2.4.4.2 Carbon Dioxide Uses in Cosmetics

In the cosmetic industry, CO_2 is often used as a superior extraction medium for pure, natural fragrances and active ingredients. One may highlight the stabiliza-tion of natural pigments, ensuring premium natural and sustainable pigments with a long and stable shelf-life, as well as micronization of fat powders. In fact, CO_2 micronization technology provides a simplified and environmentally friendly alterna-tive to traditional energy-intensive processes. CO_2 micronization allows the creation of three-dimensional porous particles from fats and waxes, thereby delivering an array of textures. Thus, active ingredients can be added simultaneously generating improved functionality and structure. The simplified formulations, along with lower energy consumption, led to significant cost-savings.

Actually, in a broader sense, CO_2 may be used as an effective agent for microen-capsulation processes. Microencapsulation processes based on carbon dioxide are used to deliver, protect, stabilize, or control the release of a variety of active ingre-dients such as pigments, antioxidants, vitamins, minerals, peptides, proteins, and fragrances for the next generation of cosmetics products. High encapsulation efficiencies, for the formation of defect-free particles, are enabled by interactions at molecular level between CO_2 and compounds in the matrix. Then, CO_2 can be removed by simply depressurization, ensuring well-protected and solvent-free ingredients. Therefore, these CO_2 processes result to be flexible, scalable, and cost-effective means for tuning the processing conditions to achieve a greater control on the particle characteristics. Particle customization for a given cosmetic application includes controlled or triggered release, ingredient stabilization, formulation compatibility, and sensorial properties [377].

2.4.4.3 CO_2 for Food-Grade Bioactive Compounds

In recent years, the demand for nutritive, functional, and healthy foods has been steadily increasing [378]. This trend has encouraged the food industry to investigate innovative technologies capable of producing ingredients with enhanced functional and physicochemical properties. Through the years, numerous technologies have been developed. Spray drying is the most widely used encapsulation technique in the food industry [379, 380]. About 80–90% of encapsulated products present on the market are produced via spray drying.

Regarding novel technologies, one of the most promising is the encapsulation based on supercritical fluids [381]. Indeed, recent published literature has revealed that supercritical fluids can be an excellent alternative technology for encapsulating active compounds. Many bioactive compounds such as antioxidants [382], vitamins [383], pigments [384], and essential oils [385] can undergo encapsulation via supercritical fluids.

CO_2 is a rather suitable molecule for encapsulating thermo-labile compounds such as vitamins, tocopherols, or oils rich in omega-3 polyunsaturated fatty acids, thanks to the low temperature needed to turn carbon dioxide into a supercritical fluid. Hence, the most widely used solvent in encapsulation, micronization, and particle formation processes is supercritical carbon dioxide (SC-CO_2) [386, 387].

Several different SC-CO_2-based technologies have been developed over the years. Such technologies are classified into several different groups, which are based on the nature of the targeted bioactive compound (being soluble or insoluble in SC-CO_2), the nature of the carrier material, and the application of the final microencapsulated compounds. Hence, according to the role played by the SC-CO_2 in the encapsulation techniques, the CO_2 can be categorized as a solvent, an anti-solvent, a solute, a co-solvent, an extractor and anti-solvent, an atomization, or a drying medium.

For instance, should SC-CO_2 behave as a solute, this means one deals with gas-saturated solution process [388]. Nevertheless, should SC-CO_2 act as an anti-solvent, the gas anti-solvent process, a supercritical anti-solvent process, or solution-enhanced dispersion by supercritical fluid process are the technologies that must be considered [389–391].

The negative impact of the consumption of food and beverage industry on energy, water consumption, climate change, and other environmental subsystems was analyzed by means of a critical and systematic review of more than 350,000 sources of evidence and a short list of 701 studies on the subject of GHG emissions. Using a socio-technical lens that examines food supply and agriculture, manufacturing, retail and distribution, and consumption and use, the review identifies the most carbon-intensive processes in industry as well as the corresponding energy and carbon "footprints". The benefits of decarbonizing the sector were examined considering current and emerging practices, including 78 potentially transformative technologies. Benefits were identified in areas such as savings on energy, carbon, and costs; sustainability or health, including the barriers in the dimensions of financial and economic, institutional and managerial, and behavioral and consumer. Finally, ways of overcoming these barriers through funding, business models, and policies were discussed, from which a set of research gaps were identified [392].

2.4.5 Environmental Applications

Mineralization finds an additional environmental application for neutralizing acid waters from mining lakes. In fact, weathering of iron sulfide (pyrite) produced during lignite mining leads to acidification of neighboring waters. Lignite is used for firing power plants with a subsequent emission of CO_2. The fly ash produced in these plants is also a source of Ca^{2+} and Mg^{2+} that could be used to mineralize the emitted CO_2. It has been found that fresh fly ash has a sequestration potential of 33 g_{CO2}/kg_{ash}. A first application combining the fly ash with addition of calcite succeeded in the treatment of lake Burghammer, in Germany from March 2009 to December 2010 [393].

One of the most ambitious environmental applications of mineralization is the direct decarbonization of flue gases. Mineralization of flue gases to remove ambient CO_2 as carbonates is a way to contribute to mitigate its GHG effect, for this reason research in MC has moved towards processes that combine the capture of CO_2 from flue gases with its conversion into stable carbonates. However, it requires the use of highly reactive materials that are able to dissolve at low CO_2 pressures and temperatures. A study was conducted using activated serpentine to determine its carbonation potential under flue gas conditions. One-step carbonation experiments were performed in stirred reactors with gas-dip tubes, at CO_2 partial pressures up to 1 bar, temperatures between 30 and 90 °C, with and without concurrent grinding using a ball mill. The pH and solids were controlled in situ and the degree of carbonation of the products was determined by thermogravimetric analysis. Despite the low CO_2 pressure, carbonation was effective, since formation of nesquehonite and hydromagnesite, two magnesium carbonates, was confirmed. However, it is important to point out that under all conditions investigated, carbonation did not exceed 20%. It was concluded that after the start of precipitation, the reactor solution in one-step carbonation experiments reaches equilibrium conditions with respect to both serpentine dissolution and carbonate precipitation [394].

An improved economic viability together with a reduction of environmental impact of pulp mills might be achieved by their conversion into forest biorefineries for producing bioenergy and biomaterials. The recovery of lignin from process streams for further use, in a variety of innovative ecological processes, represents one of the key challenges in achieving this goal. The fundamental chemical structure of lignin recovered from Kraft pulp streams was studied using an acid wash/precipitation methodology. NMR and SEC techniques were employed to determine functional group analysis and molecular weight profiles, with promising results for future conversions, such as low hydroxyl (oxygen) contents and low molecular weights ($\sim$3000 g/mol) [395].

It is a fact that the cement sector as an end-user of the building sector gives rise to environmental applications. Even though China's economic growth has been largely based on coal consumption, the country has realized the importance of shifting its economic development away from relying on fossil-fuel use. In fact, they were committed to submitting to the Conference of the Parties 21 (COP21), their results demonstrating reductions of CO_2 emissions per unit of GDP, from 60% (2005 levels) to 65%, by 2030. Their efforts focused on the construction sector (urban residential, rural residential, and service), covering 31 regions of mainland China, and taking into account the disparities in climatic and socioeconomic indicators between the regions. It is expected that the deep decarbonization in the construction sector by mid-century will contribute to achieve their goals committed through the Paris agreement. A bottom-up cost optimization model called AIM/end-use which consists of three main databases, energy and emissions factors, technologies, and socioeconomic scenarios, was used to assess the CO_2 reduction potential of efficient technologies in the Chinese construction sector. Five scenarios were designed to illustrate emission trajectories up to 2050. The results show that when energy constraints and emission targets are introduced in the mitigation scenarios, new generation biomass greatly contributes to emission drops. The reduction potential in the near-zero emissions scenario comes mainly from the urban residential sector, and to achieve deep decarbonization by 2050, it is important to achieve a significant reduction in energy consumption per capita in addition to improving in both, urban and rural households. Results suggest that deep decarbonization policies can significantly reduce air pollutants [396]. Reduction of CO_2 emissions from industry could be accomplished by MC of construction and demolition waste. MC is a stable and safe alternative to the use of underground geological formations for CO_2 storage. This method of capturing and storing CO_2 has been considered for years as a target issue in the cement chemical research sector. The primary sources of calcium for MC for fixating CO_2 in a stable material using an inert process come from concrete, cement, or ceramic blocks. For the carbonation tests, a type of concrete was selected, which was crushed, separated into size fractions, moistened at 20%, and tested for 24–720 h at a CO_2 pressure of 10 bars. The destruction of the portlandite and ettringite phases was determined. Calcite precipitated as carbonate phase. Maximum carbonation was reached at 72 h, fixing 6.5% CO_2 [397].

2.5 Source of Energy and Energy Requirement Improvements

A novel approach to energy system archetypes was technologically advanced and it can be assessed directly, ranking similar countries regardless of geographic location. Through the development of energy system models, it permits finding possible optimal technological combinations for decarbonization strategies world-wide, allowing the analysis of a wide variety of countries. A transferable global database was set up that permits comparison of peer countries facing similar challenges by defining archetypes. Standardized modeling rules were developed, following a grouping approach with the 15 defined archetypes, and the results were validated. The variation between all countries and their corresponding archetypes improves by 44% compared to the variation between countries and their geographic subregions globally [398].

In order to accelerate the decarbonization of the energy system and protect the environment, it is necessary to carry out an energy transition from a system dominated by the burning of fossil fuels to one with zero net emissions of CO_2, the main anthropogenic greenhouse gas. The energy transition is essential to mitigate climate change, protect human health, and revitalize the economy. To help shed light on the importance of achieving a net-zero transition for the United States, a committee of experts convened by the National Academies of Sciences, Engineering, and Medicine researched and identified key technological and socioeconomic goals for how to better decarbonize its transportation, electricity, buildings, and industrial sectors that must be accomplished to place the United States on track to achieve net-zero carbon emissions by 2050. A report presents a policy plan that outlines near-term actions for the first decade (2021–2030), including ways to support communities that will be most affected by the transition [399].

Due to the strong pressure to decarbonize electrical systems, the Canadian provinces of Quebec and Ontario, as well as the northeastern states of the United States, have committed to reducing their greenhouse gas emissions by more than 70% (from 1990 emission levels). With the existence of important hydroelectric resources available, it is possible to reduce decarbonization costs through greater collaboration and integration between regional authorities. Through the analysis of the existence of a greater collaboration and integration, the impact of emission reductions, load levels, and the availability of energy technologies, a variety of scenarios was analyzed, in order to assess the benefits of regional cooperation. The results show that, for a deep decarbonization, the electricity system costs can be significantly reduced by increasing the interconnection capacity. However, since cost-savings and benefits are not allocated evenly across jurisdictions, the creation of collaborative incentives is difficult [400].

A chemical-power IES was techno-economically evaluated. Power was generated by a power plant, driven by natural gas combined cycle, with CO_2 capture. The chemical production of hydrogen was carried out in integrated MRs, in which a membrane reformer was integrated to a membrane WGS reactor. Variables such as electrical and capture efficiency, required membrane surface area, and avoidance of CO_2 cost were analyzed. It was observed that for CO_2 capture, membrane water–gas-shift is more suitable than membrane reforming. The fact that the H_2 pressure is lower in a membrane reformer compared to membrane WGS causes high investment costs, since the membrane area being high, compression of the hydrogen fuel is necessary before entering the gas turbine. An improvement in the performance of membrane reformers could be achieved by increasing the operating temperature combined with a higher feed pressure. Gas heated reforming is the preferred option for the various upstream in membrane water–gas shift reforming options, while autothermal reforming is considered the second best option [401].

In order to improve the energy efficiency and product delivery (e.g., electricity, chemically stored energy, chemicals, fuels, etc.), chemical processes have been integrated into the power generating units, when operating in a closed circuit, including it within the pipeline concept of chemical heating. In the chemical heating pipe, an endothermic reaction is carried out, using the low-C energy source (renewable or nuclear), and the energy is stored in the reaction products as chemical energy. The products obtained in the endothermic reaction are then transported to a different location, where an exothermic reconversion reaction occurs, allowing the recovery of the original reagents and energy as heat, which can be used for the generation of energy. The chemical heating pipe concept was previously presented, reviewed, and discussed, along with some examples in Ref. [94] (and references therein). This technology is used as an exothermic semi-cycle in PtG systems, methanation of CO_2 or CO. As a source of energy, renewable energies are one of the favorite choices, as evidenced by its continuous growth. Nuclear energy has also been recently receiving innovative attention [402]. Globally, the PtG program has devoted great efforts to the development of methanation process technologies, both catalytic and electrocatalytic processes [199, 222, 403, 404]. Notwithstanding, the chemical energy transmission systems (CETS) has been coined to describe these energy storage and transmission systems [405]. Integration of MDR processes to solar energy has probably received the most attention. In the case of Solchem [96, 406–408] and CLEA [97, 409, 410] processes, mentioned above, MDR is the endothermic reaction, while the methanation reaction is the exothermic other half-cycle reaction. Both cases can be considered a way of converting solar energy into chemical energy for facilitating storing and/or transmission. Other attempts for solar energy integration to MDR have been reported in Refs. [402, 405, 411–416]. The performance of a solar-chemical heat pipe was studied using MDR as a vehicle for solar energy storage and transport. The endothermic dry reforming reaction was carried out in an Inconel reactor, packed with a rhodium catalyst. The reactor was suspended in an insulated receiving box that was placed in the focal plane of the Schaeffer Solar Furnace at the Weizmann Institute of Science (Rehovot, Israel). The exothermic methanation reaction was carried out in a tubular reactor filled with the same Rh catalyst and fed with the reformer products. For both reactions, conversions obtained were superior to 80%. The products of the closed-circuit reformer and methanator were compressed and stored in separate tanks. The reforming and methanation reactions were carried out consecutively and the process was repeated for nine cycles. The results obtained are in good agreement with the expected results [417]. Further experiments were carried out to study the influence of CO-rich synthesis gas in the back exothermic methanation reaction. The best performance was obtained with a six-stage adiabatic methanator. The experimental results were compared with computer calculations, assuming equilibrium at each stage. The agreement between experimental and calculated results is satisfactory. Scale-up work on a 400 kW unit is in progress [418]. A multistage methanator loaded with the same Rh catalyst and fed with reformer products was used to run the exothermic methanation reaction. An overall satisfactory performance was observed when the integrated process ran the two consecutive reactions and repeated in a closed loop, for over 60 cycles. Scale-up work was announced to be progressing [298].

For RE systems, the power-to-gas (PTG) technology is recognized as a potential option to solve long-term problems of efficient and environmentally friendly storage. The generation of electricity through RE systems makes it possible to feed both the electricity grids when it is needed or store it when it is excessive. A thermodynamic analysis was carried out on an integrated wind energy system composed of a wind turbine, a proton exchange membrane electrolyzer, and a methanation unit. The energy and exergetic efficiencies of the developed global system are located at 44% and 45%, respectively. The methanation unit works based on the Sabatier reaction for the production of synthetic natural gas (SNG). For methanation, a selective vapor permeation membrane was considered, and the products were integrated with other parts of the system for heat recovery. It was observed that an increase in wind speed causes a decrease in exergetic efficiency and an increase in the production of hydrogen and methane. The multi-objective optimization method based on a genetic algorithm was used to determine the optimal values for the decisive variables. The results show that the exergetic efficiency of the entire system is 41% and the production of CH_4 is 1.68 kg/h, which are the optimal values in which the wind speed is 4.33 m/h and the power coefficient becomes 0.57 [419]. The expected trend development of costs related to CO_2 electrolysis and methanation was analyzed, and a projection to 2030 with a 2050 outlook was provided. The results show substantial cost reductions for electrolysis and methanation over the recent years and if the cost projection follows the current trend, a further drop in prices to less than €500/kW of electrical energy input for both technologies until 2050 is estimated. Most of the examined projects are located in Germany, Denmark, the United States of America, and Canada. Following an exponential global trend to increase installed power, current PTG applications are operated at approximately 39 MW. Investigation on equal terms of hydrogen and substitute natural gas was also analyzed for the number of projects [222]. Gas-fired power systems based on the Sabatier process are potential answers to the problems related to meeting the technical and economic challenges of mitigating global CO_2 emissions. Since the main problem of energy-to-methane storage technology is its low overall conversion efficiency, pinch analysis approach (a methodology for minimizing energy consumption and optimizing energy recovery in an industrial chemical process, minimizing capital investment) was applied to improve energy recovery which represents a key factor in increasing the overall efficiency of methane storage plant. An exergetic analysis and TEA of the proposed plant was carried out to evaluate the main sources of irreversibility and to calculate the production costs of CH_4

based on the main parameters of the plant. A Sabatier conversion yield of 93.48% was achieved, producing 0.42 kg of CH_4 for each kg of CO_2 captured with an improved cost of €53/MWh. The results of the study have revealed the great potential of this solution as an "energy storage" and CO_2 capture facility [178].

Based on the good fit of experimental and prediction, a developed computer model for closed-loop chemical heat pipe is recommended for the design of larger systems. The modeled system consisted of a solar source to drive the endothermic methane reforming coupled to the exothermic methanation of syngas [420].

Methanol is a convenient liquid fuel and building block for synthetic hydrocarbons and chemicals. Its production from CO_2 has been described in the previous section. The possibilities of producing renewable methanol (and DME) lead to propose a new economy in which these compounds replace petroleum and natural gas, respectively [157]. The various aspects of current and alternative energy sources to optimize production were discussed [153, 421], paying particular attention and emphasis on this methanol economy proposal (see Fig. 2.8). Since CO_2 is expected to be obtained from industrial effluents or from the atmosphere, this approach represents another possibility to reduce global warming worries.

The debate on whether it is a methanol or a hydrogen economy fills current scientific, technological, political, and economic discussions. Methanol has greater possibilities to become a functional replacement for fossil fuels, using existing infrastructure and being globally produced through different pathways. However, at the end of the day the final answer depends on which one can be sustainably produced.

An appropriate methodology for solar-driven photoreduction of CO_2 to methanol has been investigated. Various photoreactors were evaluated and a comparative analysis of the influence of catalyst type, irradiation time, irradiated area, light intensity, and methanol yield was executed. The effect of these parameters on the performance

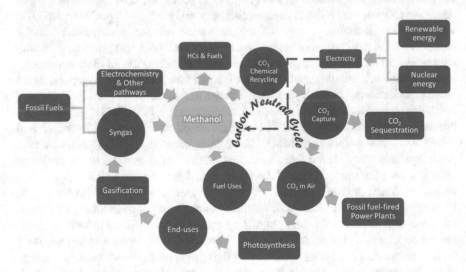

Fig. 2.8 Summary of the methanol economy

of the solar photoreactor was investigated. The main finding is that technology is far from being cost-effective and more research is needed for both, the methanol synthesis process and the solar devices. These activities have to keep in mind the ultimate objective of a larger scale process for commercial application. Aspects to be subjected to further investigation include energy losses, efficiency drop, long-term stability, and cost-effective photocatalyst [422].

At this point, the high energy requirements of all CO_2 conversion reaction have become evident. Therefore, low-carbon or neutral carbon emission sources are needed, to provide the energy demanded by any utilization or conversion process. The use of solar energy in the Solchem process [95, 96] and for the CLEA Project at Sandia National Laboratories [97] was mentioned above, as examples for enhancing the decarbonizing action of the MDR processes and placing them into the negative carbon category. Additionally, there are some other pathways for artificial carbon dioxide fixation using solar energy. Examples of these pathways for solar carbon dioxide fixation include the homogeneous photochemical reduction of CO_2, the heterogeneous photochemical reduction of CO_2, the photoelectrochemical CO_2 fixation, the electrochemical reduction of CO_2 using solar electric power, and the solar production of hydrogen followed by hydrogenation of CO_2.

Another source of renewable electricity can be non-thermal plasma (NTP) technology for CO_2 conversion, using a plasma reactor. Different types of plasma reactor configurations have been investigated for CO_2 conversion, also involving how the electricity is supplied and its power rating. The performance on CO_2 conversion was studied varying the dielectric barrier discharge, microwave plasmas, and gliding arc plasmas. All options resulted quite promising [423]. Catalytic methanation has also been examined using plasma [424].

It is imperative that decarbonization is made a priority objective to achieve climate change mitigation. Therefore, it is extremely important and necessary to follow a decarbonization strategy that includes improvements in renewable energies, even when this approach is not enough to avoid the effects of climate change. We must be aware of the real situation of energy consumption in which we find ourselves. During COP 23, the need for a transition to a future with zero-carbon emissions, necessary to avoid the negative effects of climate change, was discussed. Therefore, energy production must be transformed into almost carbon free by 2050. The experts at COP 23 meeting agreed that to achieve a zero-carbon world, we must stop using burning fuels. Current energy sources are discussed to achieve a future with zero or very low carbon emissions and the necessary support policies for renewable energies and the transition to a low-carbon economy. It is vital to think on the possibility of decarbonizing not only electricity, but also heat and fuels for transport should be considered. The objective of the analysis carried out is to perform a realistic approach on what to do taking into account the current situation of energy consumption and the objectives of reducing greenhouse gas emissions [425].

The effectiveness of the incorporation of RE into the decarbonization strategies in terms of the mass of CO_2 mitigated per unit of energy consumed has been recently reported for several strategies [426] and some of the data is presented in Table 2.4.

Table 2.4 Effectiveness of some decarbonizing strategies, measured as CO_2 mitigated per unit of RE (g_{CO2}/MJ)

Decarbonization strategy	CO_2 mitigated (g_{CO2}/MJ)
Transport sector electrification	189
Coal-fired power plants replacement	291
Natural gas power plant replacement	141
H_2 production (by water electrolysis) for use in producing methanol (for fuel applications) thermocatalytically from CO_2	51
DAC	110–185

Nuclear energy plays an important role in this decarbonizing strategy. The chemical reactions of CO_2 that can be induced by ionizing radiation were the subject of a recent review article [277]. A suggestion for H_2 production, by combining radiolysis of CO_2 with the water–gas-shift reaction, was also made [277]. Furthermore, nuclear power plants (NPPs) generate energy in different forms, i.e., radiation, heat, and electricity. Therefore, heat and electricity from NPPs can also be supplied for CO_2 conversion or utilization. Some of the opportunities for chemical processes that might benefit from integration to NPPs were identified in a recent article [94].

A standardized energy policy tool, named as the Regional Energy Hub Framework, is a grid that allows countries to generate "common interest" assets. The grid will need to meet the long-term challenge of GHG mitigation through effective optimization of the energy supply mix. This can be achieved either (i) by forming REHs in neighboring developing electricity markets (ii) or by interconnecting with other REHs. The necessary framework was applied to identify and recognize the option value of the PJM capacity market. The results obtained from the analysis show that there is an intrinsic option value between generation and interconnection assets [427].

Exergy analysis was used to investigate a concept for natural-gas-fired power plants with CO_2 capture and applied to the decarbonization of natural gas by autothermal reforming before combustion, producing a hydrogen-rich fuel. This process achieves integration between the combined cycle and the reform process. A net production of electrical energy of 47.7% of the LHV or 45.8% of the chemical exergy of the supplied natural gas was achieved. The chemical exergy of the captured CO_2 and its compression at 80 bars represented 2.1 and 2.7%, respectively, of the chemical exergy of natural gas. For a conventional combined cycle without CO_2 capture, the net electrical energy production was 58.4% of the LHV or 56.1% of the chemical exergy of the fuel. A detailed interruption of irreversibility is presented. The effect of varying supplemental combustion for reformer feed preheating in decarbonized natural gas power plant was investigated, finding that supplemental combustion increased total irreversibility and decreased net plant output. The effects of increasing the gas turbine inlet temperature and gas turbine pressure ratio were also studied: it was observed that for the conventional plant, higher pressure led to higher efficiency, in some cases. While a higher pressure ratio in the decarbonized natural gas process led to higher irreversibility and lower efficiency of the thermal plant [428].

2.6 Net-Zero Pathways

In order to set the global temperature increase in 1.5 °C, carbon emissions need to be reduced to stabilize the atmospheric CO_2 concentration. Currently, emission rates are pushing towards the undesired direction, i.e., the increase of atmospheric CO_2 concentration. Worldwide, more than 140 countries have set an aspiration to reach net-zero emissions by certain year, most typically 2050. In the "Net-Zero Initiative", emissions and removal are balanced leading to a steady atmospheric CO_2 concentration, at which global temperature rise is set at no more than 1.5 °C. Major economies are actively participating in this initiative, representing 61% of GHG emissions, 68% of gross domestic product (GDP), and 56% of the world's population [429]. It was expected that the UN Climate Summit in Glasgow would add more details on commitments, planning, targets, and metrics, which do not seem to be on the plate yet [430]. In the past, 2020 expired rules established a trading system, in which high-emitting countries were allowed to balance their emissions buying carbon credits from lower emitting countries. However, in practice, this has no net emission-reducing effect globally, and climate change remains threatened. Once again real commitments are needed as has been stated: *"a pledge that doesn't include meaningful reductions increases the risk that catastrophic climate change will become unavoidable"* [431].

In the context of net zero, neutrality refers to balance. Thus,

(a) emissions neutrality refer to the balance between emission sources and sinks,
(b) carbon neutrality refers to the total annual CO_2 anthropogenic emissions being net zero,
(c) GHG emission neutrality (also known as climate neutrality) is the sum of all Kyoto GHG emissions being net zero (in CO_2-equivalent).

In order to determine neutrality targets at the global level, scenarios from integrated assessment models (IAMs) were used to project the phase-out year to reach net zero (for the 1.5 and 2 °C targets), for different countries. The results obtained from the cost-optimal scenarios for the 1.5 °C target indicated that while Brazil, United States (CO_2 and all GHGs), and Japan (GHG only) are projected to have an earlier phase-out year than the global average, India and Indonesia would have a late phase-out year. China, the EU, and Russia were found to fall near the global average. Some of the countries would phase out at different times for CO_2 than for GHG [432].

The definition, development, implementation, adoption, and practice of net-zero strategies are gaining more and more track all over the world. Many governments are already committed to reducing their emissions. However even if these governments attained the pledged reductions that would not be enough to bound the global temperature rise to 1.5 °C. The Shared Socioeconomic Pathway (SSP) Database [433] and Integrated Assessment Modeling Consortium (IAMC) Database [434] found that by implementing technologies and measures through 2020–2050, for 50% emissions reduction per decade, nearly 90% total reduction by 2050. Therefore, further coupling with substantial carbon removal technologies would increase the probabilities of reaching levels beyond 90% and closer to the limiting rise target of 1.5 °C.

The energy and industrial sectors are the core of any net-zero strategy, centering all the attention on energy intensity and carbon footprint. The "hard to decarbonize" qualification given to industry is associated with the multiple and variable processes involved. Four decarbonizing approaches have been recommended for the industrial sector: (1) zero-carbon fuels, (2) zero-carbon heat sources, (3) electrification of heat, and (4) better heat management, particularly for mitigating emissions related to heat [435]. Additional approaches were included according to the IPCC report of Ref. [436], namely, the reduction of energy demand, improvements in energy efficiency, and deploying innovative processes and application of CCS. Once more it is important to emphasize that the strategy cannot jeopardize energy security, neither energy affordability nor economic growth. Achieving net-zero emissions requires the participation and commitment of the entire society and stakeholders in key roles through the development of cost-effective clean technologies, the definition of enabling policies and promoting incentives, the needed investments, and the behavior changes among many other factors.

The accountability or metrics of net-zero approaches is defined by the Greenhouse Gas Protocol [437]. The scope refers to both the type of GHG and the type of emission. In principle, according to Kyoto Protocol, net zero involves all GHGs though currently the focus is given to CO_2. Three scopes were recommended by IPCC as Scope 1 refers to direct company owned or controlled emissions occurring at source, Scope 2 are the emissions associated with the production of energy consumed by a company, and Scope 3 accounts indirect emissions associated with company activities from sources not owned or controlled by a company [437].

The dynamics of building sector implies addressing 2D-type concerns, one side is the current assets and the other are the future assets, as well as the side of the occupants and the owners in the other. Succeeding in decarbonizing the existing assets is considered essential, as it is the design and development of net-zero-driven approach for management. The pathway to net zero was suggested to be taken in three steps: (i) lifecycle-driven planning that maximized emissions reduction at the minimal cost and operational disruption; (ii) adapting typical refurbishment design briefs; and (iii) carbon reduction projects [438]. In the UK, an approach was established to determine the required energy demand reduction to meet 2050 net-zero goal (the economy fully powered by zero-carbon energy). The "Paris Proof" methodology indicated the needs for very steep cuts in energy demand, which they applied to office buildings, indicating the need for redefining the office designs, construction, and operation, including the monitoring of energy performance metrics [439]. Buildings consume 46% of energy in the form of heat though industrial heat is in fact the largest consumer. Globally, heat represents half of the energy demand, which is typically met by fossil fuels and translate into 40% of the global CO_2 emissions [440]. A comparison of heating options (natural gas boilers, air source heat pumps, hydrogen boilers, and direct electric heaters) positioned heat pumps as the top to offer the most environmentally sustainable option [441]. Three approaches for net-zero energy in residential buildings were suggested and analyzed: energy infrastructure connections, RE sources, and energy efficiency measures. A published work reviewed the technology options for addressing each of these approaches and found that although there is not a single net-zero energy building configuration that is optimal for all climates, regulations, building codes, and markets, there are available technologies to adapt and adopt a configuration for a given situation. Urbanists, planners, designers, and/or developers could select a convenient suite of technology and building parameters to adapt to specific conditions and requirements [442].

Cement is used for making concrete, which is the globally largest volume of any manufactured material. Thus, cement is directly related to the building sector, the cement industry is responsible for 7% of the global CO_2 emissions. This situation might become aggravated by economy and population growth driving an increase in the demand for cement of 12–23%, by 2050 [329]. The industrial contribution to US emissions grew from 16% in 2016 to 23% in 2020 [443], indicating that the decarbonization efforts of this sector are behind those of the others. The IEA report on the cement industry recommends three direct actionable approaches and two longer term strategies for achieving 2050 net-zero goals, these are improvements on energy efficiency, switch to alternative energy resources, and reduction of the clinker-to-cement ratio, as well as development and deployment of emerging and innovative technologies and seek for alternative (decarbonized) binding materials for functional replacements of cements [329]. A study using a fully integrated TEA/LCA methodology that involves the definition of one common goal and scope, the use of one common inventory model, and the calculation of combined indicators provided evidence that none of the single approaches recommended by IEA can lead to 2050 goals. For instance, MC could result in 11–16% CO_2 emission reduction while combining carbonation with replacing fossil resources with RE leads to 45% in emission reduction. Improvements in energy efficiency resulted in 14% reduction in the production costs of net-zero cement [444].

The electrification of the transport sector is a common path considered to achieve net-zero objectives. A study reports the impact of banning the sales of internal combustion engine cars on the decarbonization targets in Sweden. Obviously, reductions of tailpipe CO_2 emissions decrease carbon footprints; however, a 2030 ban will not be enough to achieve 2050 net-zero goals. An earlier ban accompanied by an increased use of biofuels was suggested [445]. Currently, electrification of the long-distance transport sector appears so difficult and far from implementation that drop-in fuels emerge as an effective and more probable decarbonizing option. The low-carbon fuels standard (LCFS) is an example of policies, to create incentives to fuel manufacturers. These policies should cover the space of the manufacturing processes (carbon intensity) as well as the incorporation of biofuels or renewable fuels (renewable content), in the final product. Co-processing is an alternative that addresses both aspects simultaneously. However, today, standardized accountability methods and terminology are deficient. A multiple-regression-based mass balance approach (based on observed yield) combined with ^{14}C analysis has been proposed to determine the renewable content in final fuels, produced by co-processing [446]. Among transportation vehicles, aircrafts represent the most challenging devices.

Three approaches have been visualized for this industry, namely, electrification, alternative fuels, and carbon offsetting, e.g., emission removal from the atmosphere or carbon trading [447]. DAC and conversion of the captured CO_2 into jet fuel is an example of this type of pathways. However, this CCU pathway is not enough to achieve net-zero goals and more CCUS approaches will be needed [447]. Moreover, as it will become evident next, the challenges in aviation are complicated enough to require other approaches. The electrification of certain onboard systems (e.g., de-icing, pressurized devices, etc.) is already available and in practice though their contribution to emission reductions is marginal. Current electrification technology solutions can only be considered for short distance flights of about 1 h long. Thus, the aviation sector is considering a combination of electrification and alternative low-carbon/renewable fuels as the pathway towards net zero [448]. The growth in travel and freight demand in commercial flights would lead to triplicate carbon emissions by 2050. An estimated 85% emission reduction could be achieved through demand management, improvements in energy efficiency, and incremental incorporation of low-carbon/renewable fuels. ICCT shows how to design pragmatic policies using research on aircraft technology, fuel efficiency, and alternative fuels. Thus, under this scenario, the net-zero phase-out year does not seem to be possible by 2050 [449].

The clean technologies of choice for development are those exhibiting negative emissions (NETs). Any emerging NET has to be assessed within the complex and comprehensive framework in which it would perform. The NETs critical performance aspects (e.g., feasibility, effectiveness, and side impacts) have to be considered holistically. Since there is no universal solution, each situation could be solved most likely with an optimal configuration, constituted by a NETs portfolio. A multi-criteria decision analysis method applied to the most promising NETs led to the conclusion that only an optimized technology portfolio timely deployed will sustainably contribute to mitigate climate change. Unoptimized solutions can be unfeasible, useless, and their collateral impacts can even create larger problems. Seven of the most promising NETs were evaluated: afforestation and reforestation (A/R), biochar (BC), bioenergy with carbon capture and storage (BECCS), direct air carbon capture and storage (DACCS), enhanced weathering on land and in oceans (EW), ocean fertilization (OF), and soil carbon sequestration (SCS) [450]. A comparative analysis was performed on energy systems, BECCS and direct air capture (DAC) with geological carbon storage (DACCS), using a bottom-up energy system model TIMES-Japan. They found that it was possible to meet long-run climate goals with an earlier deployment of BECCS, using domestic biomass. DACCS can become a supporting and supplementary technology at a later period. Finally, this study also showed that an earlier deployment of DAC systems combined with CO_2 utilization in fuel production results in cost-effective NETs [451]. BECCS's massive deployment by the end of the century would be unsustainable, and A/R alone cannot be considered an effective climate change mitigation solution [450].

The 2018 IPCC Special report calls for preservation of land carbon stocks, e.g., through reduced deforestation and afforestation [436]. The successful transformation of the agriculture and forestry, by practicing new management measures and deploying in large bioenergy production, is the main target in the path to net zero of the land sector.[1] These two approaches lead to yearly emission reduction of 30% (~15 $Gton_{CO2e}$/y) [452].

Thus, the key role of bioenergy during and beyond the energy transition has been established throughout multiple analyses. Typically, long-term planning of policies is supported by analyses outputs. An optimization exercise based on a techno-economic model for the evaluation of cross-sectoral bioenergy policies, under high spatial resolution and long-term temporal resolution, has been developed to describe systems in the two-dimensional space–time scenario. The model showed that optimizing energy cost and decarbonization claims for multi-sectoral collaboration between all energy sectors (power, heat, and transport). A reasonable transition requires bioenergy to replace 30% of power, 27% of heat, and 12% of transport energy demands. The analysis also indicated that in the power and transport sectors, more effective reductions of CO_2 emissions can be targeted through new policies prioritizing bioenergy than those achieved through CO_2 reduction policies alone [453]. In Portugal, the impact of biopower and biofuels was evaluated using a bottom-up linear optimization energy system, the integrated MARKAL-EFOM system (TIMES) model. Regardless of the benefits that bioenergy might bring about, its incorporation in the energy menu will not be enough to achieve 2050 net-zero goals and the increases in particulate matter emissions have to be brought into the analysis [454]. In order to assess the economic impacts of the net-zero emission target in New Zealand, an integrated forest-computable general equilibrium model was developed and used to study the carbon trading market. They compare the possibility or not of imposing a legally binding target to the agricultural sector. It was estimated a cost of equilibrium carbon permit of US$60/$ton_{CO2e}$ or US$48/$ton_{CO2e}$ when the agricultural sector is bound and when it is not, respectively. The impact on the GDP was of 1.4% losses for 22% emission reduction upon incorporation and only 1.5% GDP losses for 5% emission reduction, without binding agriculture. It was suggested a proactive definition and implementation of policies that could leverage the cost-effectiveness of the mitigating technologies, in order to balance the compliance costs with the economic impacts [455].

An analysis tries to establish whether there would be business interests or incentives (both economic and environmental) in their company's practices for reducing carbon footprint. This study targeted the understanding of the dynamics of carbon reduction initiatives and their emission and financial consequences, by reviewing data available in the published literature. The data gaps were filled by analyzing the dynamics of emission prevention, of product stewardship and the sustainable development perspectives through operational improvements, stakeholder integration, and a shared vision. They found that the carbon reduction practices may not be effective nor rewarding when they are reactive or symbolic. Therefore, reported carbon reduction efforts do not seem to translate into positive environmental and economic outcomes [456].

---

[1] The land sector refers to the agriculture, forestry, and other land uses, involved in activities leading to anthropogenic emissions.

At this point, it becomes evident that more than one technical pathway is needed but also supporting and incentivizing policies have to be defined and implemented. The multidimensional nature and the holistic focus of the policies have been widely reported in the literature, some examples will be given in the next chapter and as a matter of exemplification the reader is referred to Ref. [457]. In these articles, the authors formulate the problem as a chance-constrained equilibrium with equilibrium constraints that consider multiple state regulators acting in coordination with in-state power companies to implement Renewable Portfolio Standard goals in the least cost manner.

In defining their path to net zero, the UK government defined and adopted a set of principles to underpin its actionable plan [458]. The top 20 principles were

1. Informing and educating everyone	2. Use of mix of natural and technological solutions
3. Support for sustainable growth	4. Fairness for the most vulnerable globally
5. Urgency	6. Regular independent checks on progress
7. Long-term planning and a phased transition	8. Making the most of potential benefits for everyone
9. A joined-up approach across the system and all levels of society	10. Underpinned by scientific evidence and focused on the big wins
11. Ensuring solutions are future-proofed and sustainable for the future	12. Think about our impact globally and be a global leader
13. Achievable	14. Equality of responsibility
15. Protecting and restoring the natural world	16. Local community engagement
17. Leadership from government	18. Transparency and honesty
19. Fairness within the UK	20. Everyone should have a voice

From these principles, in the UK the center is society, and everybody is responsible (both personally and professionally) for the transit and achievement of net zero. An assembly was structured to discuss the areas of attention, namely, travel on land, travel by air, at home, eating and land-use, purchases (products, uses, applications, and manufacturing), electricity sources, and GHG removal. Detailed recommendations were given for each of these areas and can be seen in Ref. [458]. As a matter of example, for the passenger cars' transportation, three approaches were given: banning the sale of cars using fossil fuel by 2030–2035; reduction in the usage of cars, by an average of 2–5% per decade; and improvements on public transportation. Similarly, in the area of purchases recommendations included manufacturing using less and lower carbon energy and materials; more repairing products and devices; more sharing of products, devices, and services; enhanced and larger deployment of education on choices and changes in individual behavior; increase, improvements, and facilitating recycling (including labeling products on their associated C-emissions); and long-term commitment from government and parliament (political decoupling) [458].

Regarding the technical approaches, a single pathway, e.g., CCS, C-footprint reduction, CU, electric cars, bioenergy, etc. cannot stand alone to fulfill the 2050 net-zero goals. The Fondazione Eni Enrico Mattei (FEEM) and the Sustainable Development Solutions Network (SDSN) have published a report [459], in which six decarbonization pillars have been identified as guidelines for countries to develop their roadmap to achieve decarbonization by mid-century. This report was the result of the analysis of the status of R&D and technology in four carbon-intensive sectors: power, industry, transport, and buildings. Although land-use and agriculture were identified as critical additional sectors, their analyses were not included. Additionally, emphasis was noticed on the inability of any single policy or technology to achieve decarbonization by itself or be implemented without due consideration to the interplaying consequences in the entire complex system. These six pillars were.

1. Zero-carbon electricity.
2. Electrification of end-uses, particularly for end-user sectors based on fossil-fuel energy.
3. Green synthetic fuels, including the development and deployment of a wide range of synthetic fuels, adapted to carbon-intense sectors.
4. Smart power grids consider systems capable to shift among multiple sources of power generation and various end-uses, without detriment on efficiency, reliability, and cost.
5. Material efficiency, warranting choices, and flows to continuous efficiency improvement and minimize waste.
6. Sustainable land-use, involving mainly the agriculture sector, to minimize GHG emissions from deforestation, industrial fertilizers, livestock, and direct and indirect fossil-fuel uses.

In a way, Greig re-combined these six pillars into five foundations, for underpinning the net-zero pathway [460]. These five foundations were.

1. Increase energy productivity (energy consumed per unit of GDP): improve energy efficiency, develop less energy-intense processes, and change practices and behavior.
2. Decarbonize electricity supplies by replacing fossil-fired plants with wind, solar, biomass, hydro, and/or nuclear. These power technologies are land-intense and the sustainable use of the land (the 6th pillar from SDSN [459]) need to be addressed within this Greig's foundation.
3. End-use electrification, particularly the transportation sector, household heating, and process heat. This foundation imposes more requirements and a stringent sense of urgency for foundations 1 and 2.
4. Decarbonize fuels and energy carriers, in this foundation hydrogen, biofuels and biomass play the protagonist roles.
5. Implement and deploy CCUS, installed capacity and processes would need CCUS for transitioning effectively and smoothly towards net-zero goals.

As mentioned above, the industrial bulk of energy end-uses is related to heat and steam. Among the end-uses, low-temperature heat, mechanical work, and cooling are comparatively readily to electrify, in opposition to high-temperature heat. Since heat pumps cannot generate it, electricity most of the time results prohibitively expensive

[301]. The 1.5 °C 2050 goal of the net-zero strategy imposes significant transformations of energy-intense and carbon-intense sectors, i.e., transportation, energy,
industry, and land and more importantly substantial deployment of C-removing and
C-management (e.g., CCUS) technologies. The transformation suggested for energy
systems includes fossil-fuel phase out, energy efficiency improvement, low-C or
net-zero-C electricity, industry, buildings and transport, and extensive deployment
of CCUS technologies [436].

The criticality of the land-use and agriculture sectors has merited in-depth analysis
and definition of special policies, in pro of safeguarding food chain and water accessibility. The 1.5 °C pathway for the land sector requires 85% emission reduction
by 2050. The scenarios for successfully transiting through this pathway are characterized by a rapid shift away from fossil-fuel-fired power supply towards low-C
energy generation, reduced energy use, and carbon dioxide removal (CDR) [433].
A large gap is currently observed in the achieved progress. The political inertia,
weak governance, and poor investment have been identified as the major barriers
for delivering results. However, these barriers seem to be justified by the negative
trade-offs on food security, economic returns, and adverse impacts on smallholders.
Social measures to change the demand (e.g., reducing food waste and shifting diets)
underpinned by behavior changes have not received enough support. The increasing
knowledge gap is partially responsible for the situation and it is necessary to build
new connections bridging stakeholders from the scientific community and those in
the political affairs community. Research areas key to achieve the net-zero goals
include not only NETs (e.g., CCS, advanced biofuels) but also breakthrough technologies and approaches in behavioral science, meat substitutes, livestock production
systems (e.g., new feed), peatlands restoration, improved fertilizers, seed varieties,
etc. [452, 461]. The suggested substantial transformation of the land sector involves
a pragmatic change of forest management including a decline in deforestation, a
significant increase in afforestation and in reforestation (A/R), and reduced agricultural emissions after 2030–2040, enhancements on crop production efficiencies and
yields [462]. However, the agricultural intensification combined with forest restoration will hold considerable potential if accompanied by stringent land policies and
enforcement and demand-side measures (as mentioned above) [463]. The pathways
to 1.5 °C rely on more CDR annually than the 2 °C pathway scenarios, primarily
from BECCS, but also A/R, and CCS of fossil fuels [464].

In summary and as it has been concluded in a report from McKinsey & Co., "Net-
zero emissions can be achieved only through a universal transformation of energy
and land-use systems" [308]. Finally, the integration of process technologies within
cyclic business models is key for advancing into a sustainable decarbonized future.
This will be the topic covered in the next chapter.

References

1. P.C. Psarras et al., Carbon capture and utilization in the industrial sector. Environ. Sci. Technol. **51**(19), 11440–11449 (2017). https://doi.org/10.1021/acs.est.7b01723
2. J. Rissman et al., Technologies and policies to decarbonize global industry: review and assessment of mitigation drivers through 2070. Appl. Energy **266**(114848), 34 (2020). https://doi.org/10.1016/j.apenergy.2020.114848
3. B.P. Spigarelli, S.K. Kawatra, Opportunities and challenges in carbon dioxide capture. J. CO2 Util. **1**, 69–87 (2013). https://doi.org/10.1016/j.jcou.2013.03.002
4. S. Saeidi et al., Hydrogenation of CO2 to value-added products—a review and potential future developments. J. CO2 Util. **5**, 66–81 (2014). https://doi.org/10.1016/j.jcou.2013.12.005
5. M. De Falco et al., Hydrogen production by solar steam methane reforming with molten salts as energy carriers: experimental and modelling analysis. Int. J. Hydrog. Energy **46**(18), 10682–10696 (2021). https://doi.org/10.1016/j.ijhydene.2020.12.172
6. A. Rafiee et al., Trends in CO2 conversion and utilization: a review from process systems perspective. J. Environ. Chem. Eng. **6**(5), 5771–5794 (2018). https://doi.org/10.1016/j.jece.2018.08.065
7. R. Jevtic, Stable fire extinguishing installations with CO2 fire extinguishers. Tehnika **75**(4), 527–533 (2020). https://doi.org/10.5937/tehnika2004527J
8. L. Woźniak et al., The application of supercritical carbon dioxide and ethanol for the extraction of phenolic compounds from chokeberry pomace. Appl. Sci. (Switzerland). **7**(4) (2017). https://doi.org/10.3390/app7040322
9. K. Khosravi-Darani, E. Vasheghani-Farahani, Application of supercritical fluid extraction in biotechnology. Crit. Rev. Biotechnol. **25**(4), 231–242 (2005). https://doi.org/10.1080/07388550500354841
10. M. Raventós et al., Application and possibilities of supercritical CO2 extraction in food processing industry: an overview. Food Sci. Technol. Int. **8**(5), 269–284 (2002). https://doi.org/10.1177/1082013202008005451
11. A. Biszczanik et al., Experimental investigation on the effect of dry ice compression on the poisson ratio. Materials. **15**(4) (2022). https://doi.org/10.3390/ma15041555
12. A. Fernandez Pales et al., The role of CO2 storage IEA (Paris, France, 2019), pp. 105. https://www.iea.org/reports/the-role-of-co2-storage
13. D.E. Allen et al., Modeling carbon dioxide sequestration in saline aquifers: significance of elevated pressures and salinities. Fuel Process. Technol. **86**(14–15), 1569–1580 (2005). https://doi.org/10.1016/j.fuproc.2005.01.004
14. S. Bachu, J.J. Adams, Sequestration of CO2 in geological media in response to climate change: capacity of deep saline aquifers to sequester CO2 in solution. Energy Convers. Manage. **44**(20), 3151–3175 (2003). https://doi.org/10.1016/S0196-8904(03)00101-8
15. B. Metz et al., Carbon dioxide capture and storage. Special Report. Intergovernmental Panel on Climate Change (IPCC) (Cambridge University Press, UK, 2005), p. 442
16. D. Gielen et al., *Prospects for CO2 Capture and Storage* (International Energy Agency, Paris, France, 2006), p. 251
17. S.M. Al-Fattah et al., *Carbon Capture and Storage. Technologies, Policies, Economics, and Implementation Strategies* (CRC Press, Brussels, Belgium, 2011), p. 368
18. M.L. Godec, Global technology roadmap for CCS in industry. Sectoral assessment CO2 enhanced oil recovery. Final Report (Advanced Resources International, Inc. Arlington, VA. USA, 2011), p. 47
19. A. Esken et al., CCS global. Prospects of carbon capture and storage technologies (CCS) in emerging economies. GIZ-PN 2009.9022.6 Final Report (Wuppertal Institute, Wuppertal, Germany, 2012), p. 92
20. T. Mikunda et al., CO2 capture and storage (CCS) in energy-intensive industries. Zero Emissions Platform (ZEP). (2013), p. 38
21. S. Mcculloch et al., *20 Years of Carbon Capture and Storage. Accelerating Future Deployment* (International Energy Agency, Paris, France, 2016), p. 112

22. Clean Energy Technologies, *Canada's CO2 Capture and Storage Technology Roadmap* (CANMET Energy Technology Centre, 2006), p. 89. www.co2trm.gc.ca.
23. P. Fennell, B. Anthony, *Calcium and Chemical Looping Technology for Power Generation and Carbon Dioxide (CO2) Capture* (Woodhead Publishing, Cambridge, UK, 2015), p. 446
24. P.-C. Chiang, S.-Y. Pan, *Carbon Dioxide Mineralization and Utilization* (Springer, Singapore, 2017), p. 451. https://doi.org/10.1007/978-981-10-3268-4
25. W.J.J. Huijgen, R.N.J. Comans, *Carbon dioxide sequestration by mineral carbonation: literature review.* ECN-C--03–016 Report (Energy Research Centre of the Netherlands, 2003), p. 53
26. Y. Zhang, Progress in carbon dioxide sequestration by mineral carbonation. CIESC J **58**(1), 1 (2007)
27. A. Sanna et al., A review of mineral carbonation technologies to sequester CO2. Chem. Soc. Rev. **43**(23), 8049–8080 (2014). https://doi.org/10.1039/c4cs00035h
28. M. Delgado Torróntegui, *Assessing the mineral carbonation science and technology.* MSc Thesis from Swiss Federal Institute of Technology, ETH, Institute of Process Engineering, 2010, p. 55
29. V. Romanov et al., Mineralization of carbon dioxide: a literature review. ChemBioEng Rev. **2**(4), 231–256 (2015). https://doi.org/10.1002/cben.201500002
30. M.D. Aminu et al., A review of developments in carbon dioxide storage. Appl. Energy **208**, 1389–1419 (2017). https://doi.org/10.1016/j.apenergy.2017.09.015
31. B. Wang et al., A review of carbon dioxide sequestration by mineral carbonation of industrial byproduct gypsum. J. Clean. Prod. **302**(126930), 29 (2021). https://doi.org/10.1016/j.jclepro.2021.126930
32. C. Dessert et al., Basalt weathering laws and the impact of basalt weathering on the global carbon cycle. Chem. Geol. **202**(3–4), 257–273 (2003). https://doi.org/10.1016/j.chemgeo.2002.10.001
33. A. Bocin Dumitriu et al., *Carbon Capture and Utilisation Workshop: Background and Proceedings* (JRC Publications, Luxembourg, 2013), p. 77. https://publications.jrc.ec.europa.eu/repository/handle/JRC86324
34. I.S. Romão et al., CO2 sequestration with serpentinite and metaperidotite from northeast portugal. Miner. Eng. **94**, 104–114 (2016). https://doi.org/10.1016/j.mineng.2016.05.009
35. R.M. Santos et al., Integrated mineral carbonation reactor technology for sustainable carbon dioxide sequestration: 'CO2 energy reactor', in *Proceedings of the 11th International Conference on Greenhouse Gas Control Technologies, GHGT 2012*, vol. 37 (Elsevier Ltd., Kyoto, 2013), pp. 5884–5891. https://doi.org/10.1016/j.egypro.2013.06.513
36. K.S. Lackner et al., Carbon dioxide disposal in carbonate minerals. Energy **20**(11), 1153–1170 (1995). https://doi.org/10.1016/0360-5442(95)00071-N
37. Z.Y. Chen et al., Chemistry of aqueous mineral carbonation for carbon sequestration and explanation of experimental results. Environ. Prog. **25**(2), 161–166 (2006). https://doi.org/10.1002/ep.10127
38. W.K. O'Connor et al., Energy and economic evaluation of ex situ aqueous mineral carbonation, in *Greenhouse Gas Control Technologies* (Elsevier Ltd. 2005, 2011–2015). https://doi.org/10.1016/B978-008044704-9/50261-5
39. S.J. Gerdemann et al., Ex situ aqueous mineral carbonation. Environ. Sci. Technol. **41**(7), 2587–2593 (2007). https://doi.org/10.1021/es0619253
40. E. Benhelal et al., Direct aqueous carbonation of heat activated serpentine: discovery of undesirable side reactions reducing process efficiency. Appl. Energy **242**, 1369–1382 (2019). https://doi.org/10.1016/j.apenergy.2019.03.170
41. P. Styring, Why is CCU an important technology option for europe? SETIS Mag. **11**, 22–23 (2016)
42. R. Zevenhoven et al., Carbon dioxide mineralisation and integration with flue gas desulphurisation applied to a modern coal-fired power plant, in *Proceedings of the 25th International Conference on Efficiency, Cost, Optimization and Simulation of Energy Conversion Systems and Processes, ECOS*, vol. 6 (Perugia, Italy, Aabo Akademi University, 2012), pp. 83–102

43. F.J. Doucet, Scoping study on CO2 mineralization technologies. CGS- 2011–007 Report (CGS, Pretoria, South Africa, 2001), p. 88. https://www.academia.edu/4061042/Scoping_s tudy_on_CO2_mineralization_technologies

44. R. Zevenhoven et al., Mineralisation of CO2 using serpentinite rock—towards industrial application, in *Proceedings of the IEAGHG/IETS Iron and Steel Industry CCUS & Process Integration Workshop* (Tokyo, Japan 2013), p. 26. https://ieaghg.org/docs/General_Docs/Iron% 20and%20Steel%202%20Secured%20presentations/3_1520%20Mikko%20Helle.pdf

45. A. Hemmati et al., Process optimization for mineral carbonation in aqueous phase. Int. J. Miner. Process. **130**, 20–27 (2014). https://doi.org/10.1016/j.minpro.2014.05.007

46. R.D. Balucan et al., Energy cost of heat activating serpentinites for CO2 storage by mineralisation. Int. J. Greenh Gas Control **17**, 225–239 (2013). https://doi.org/10.1016/j.ijggc.2013.05.004

47. I.S.S. Romão, Production of magnesium carbonates from serpentinites for CO2 mineral sequestration- optimisation towards industrial application. Ph.D Thesis from Chemical Engineering, Faculty of Science and Technology, 2015, p. 188

48. A. Zimmermann et al., Carbon capture and storage by mineralisation. Stage Gate 1 Report (Energy Technologies Institute. Loughborough, UK, 2011), p. 350

49. M. Priestnall, Decarbonising flue gas using CO2 mineralisation—project experience on ships, in *Proceedings of the Keeping the Momentum* (Geological Society, London, UK, 2014), p. 9

50. M. Priestnall, Silica, metals, Mg/ca oxides, CCS (& electricity) from minerals and wastes, in *Proceedings of the Mineralisation Cluster Workshop* (London, UK, 2012), p. 5

51. M. Priestnall, *Method and system of sequestrating carbon dioxide* Patent No. GB2515995 (Cambridge Carbon Capture Ltd, 2015), 14 Jan 2015

52. M. Priestnall, Method and system of activation of mineral silicate minerals, Patent No. US9963351 (Also published as CN106573197, DK3129125, EP3129125, ES2824676, GB2516141, PL3129125, US2017029284, WO2015154887). (Cambridge Carbon Capture Ltd, 2018), 08 May 2018

53. R. Zevenhoven, Metals production, CO2 mineralization and LCA. Metals **10**(3), 16 (2020). https://doi.org/10.3390/met10030342

54. M. Priestnall, CO2 mineralisation via fuel cells. How to make CCS profitable, in *Proceedings of the Finding Petroleum CCS Forum* (Geological Society, London, UK, 2010), p. 14. http://www.findingpetroleum.com/files/event15/ccc.pdf

55. M. Priestnall, Can mineral carbonation be used for industrial carbon dioxide sequestration? Environ. Chem. Group Bull. 3 (2014). https://www.envchemgroup.com/can-mineral-carbon ation-be-used-for-industrial-co2-sequestration.html

56. R. Zevenhoven et al., A comparison of CO2 mineral sequestration processes involving a dry or wet carbonation step. Energy **117**, 604–611 (2016). https://doi.org/10.1016/j.energy.2016.05.066

57. M. Priestnall, Making money from mineralisation of CO2. Carbon Capture J. 3 (2013). https://www.carboncapturejournal.com/news/making-money-from-mineralisation-of-co2/3251.aspx?Category=featured

58. K.J. Lamb et al., Capacitance-assisted sustainable electrochemical carbon dioxide mineralisation. Chemsuschem **11**(1), 137–148 (2018). https://doi.org/10.1002/cssc.201702087

59. M.R. Dowsett et al., Exploring the scope of capacitance-assisted electrochemical carbon dioxide capture. Dalton Trans. **47**(31), 10447–10452 (2018). https://doi.org/10.1039/C8D T01783B

60. Q. Zhuang et al., From ammonium bicarbonate fertilizer production process to power plant CO2 capture. Int. J. Greenh Gas Control **10**, 56–63 (2012). https://doi.org/10.1016/j.ijggc.2012.05.019

61. A. Sanna et al., Alternative regeneration of chemicals employed in mineral carbonation towards technology cost reduction. Chem. Eng. J. **306**, 1049–1057 (2016). https://doi.org/10.1016/j.cej.2016.08.039

62. J.K. Fink, *Reactive Polymers Fundamentals and Applications: A Concise Guide to Industrial Polymers*, 2nd edn. (Elsevier Inc, 2013), pp. 1–535. https://doi.org/10.1016/C2012-0-02516-1

63. S.W. Bae et al., NO removal by reducing agents and additives in the selective non-catalytic reduction (sncr) process. Chemosphere **65**(1), 170–175 (2006). https://doi.org/10.1016/j.chemosphere.2006.02.040

64. J.H. Baik et al., Control of NOx emissions from diesel engine by selective catalytic reduction (SCR) with urea, in *6th Congress on Catalysis and Automotive Pollution Control*, vol. **30–31**. (2004), pp. 37–42. https://doi.org/10.1023/b:toca.0000029725.88068.97

65. A.D. Katsambas et al., *European Handbook of Dermatological Treatments*, 3rd edn. (Springer, Berlin Heidelberg, 2015), p. 1579. https://doi.org/10.1007/978-3-662-45139-7

66. J. Meessen, Urea synthesis. Chem. Ing. Tec. **86**(12), 2180–2189 (2014). https://doi.org/10.1002/cite.201400064

67. Y. Manaka et al., Organic bases catalyze the synthesis of urea from ammonium salts derived from recovered environmental ammonia. Sci. Rep. **10** (1), 2834, 8 (2020). https://doi.org/10.1038/s41598-020-59795-6

68. M. Poliakoff et al., The twelve principles of CO2 chemistry. Faraday Discuss. **183**, 9–17 (2015). https://doi.org/10.1039/c5fd90078f

69. BP, *Statistical Review of World Energy* (London, UK, 2020), p. 68. https://www.bp.com/en/global/corporate/energy-economics/statistical-review-of-world-energy.html.

70. M.E. Dry, The Fischer-Tropsch process: 1950–2000. Fischer-Tropsch Synthesis on the Eve of the XXI century (CATSA) **71**(3–4), 227–241 (2002). https://doi.org/10.1016/S0920-5861(01)00453-9

71. O.O. James et al., Increasing carbon utilization in Fischer-Tropsch synthesis using H2-deficient or CO2-rich syngas feeds. Fuel Process. Technol. **91**(2), 136–144 (2010). https://doi.org/10.1016/j.fuproc.2009.09.017

72. D. Kitchen, A. Pinto, LCA technology for the modernization of existing plants, in *Proceedings of the 35th Annual Ammonia Symposium*, vol. 31 (Publ by AIChE, New York, NY, United States. San Diego, CA, USA 1991), pp. 219–226.

73. J.R. Rostrup-Nielsen, Aspects of CO2-reforming of methane, in *Studies in Surface Science and Catalysis* (1994), pp. 25–41. https://doi.org/10.1016/S0167-2991(08)63847-1

74. E. Iglesia, Design, synthesis, and use of cobalt-based Fischer-Tropsch synthesis catalysts. Appl. Catal. A **161**(1–2), 59–78 (1997). https://doi.org/10.1016/S0926-860X(97)00186-5

75. M. Bjørgen et al., Methanol to gasoline over zeolite h-ZSM-5: improved catalyst performance by treatment with NaOH. Appl. Catal. A **345**(1), 43–50 (2008). https://doi.org/10.1016/j.apcata.2008.04.020

76. C. Du et al., Efficient and new production methods of chemicals and liquid fuels by carbon monoxide hydrogenation. ACS Omega **5**(1), 49–56 (2020). https://doi.org/10.1021/acsomega.9b03577

77. J.-M. Lavoie, Review on dry reforming of methane, a potentially more environmentally-friendly approach to the increasing natural gas exploitation. Front. Chem. **2**(81) (2014). https://doi.org/10.3389/fchem.2014.00081

78. Y.-H. Wang et al., Durable Ni/MgO catalysts for CO2 reforming of methane: activity and metal–support interaction. J. Mol. Catal. A: Chem. **299**(1), 44–52 (2009). https://doi.org/10.1016/j.molcata.2008.09.025

79. D.K. Kim et al., Mechanistic study of the unusual catalytic properties of a new nice mixed oxide for the CO2 reforming of methane. J. Catal. **247**(1), 101–111 (2007). https://doi.org/10.1016/j.jcat.2007.01.011

80. D. Liu et al., A comparative study on catalyst deactivation of nickel and cobalt incorporated MCM-41 catalysts modified by platinum in methane reforming with carbon dioxide. Catal. Today **154**(3), 229–236 (2010). https://doi.org/10.1016/j.cattod.2010.03.054

81. M.K. Nikoo, N.A.S. Amin Thermodynamic analysis of carbon dioxide reforming of methane in view of solid carbon formation. Fuel Process. Technol. **92**(3), 678–691 (2011). https://doi.org/10.1016/j.fuproc.2010.11.027

82. J.M. Ginsburg et al., Coke formation over a nickel catalyst under methane dry reforming conditions: thermodynamic and kinetic models. Ind. Eng. Chem. Res. **44**(14), 4846–4854 (2005). https://doi.org/10.1021/ie0496333

83. J. Gao et al., Methane autothermal reforming with CO2 and O2 to synthesis gas at the boundary between Ni and ZrO2. Int. J. Hydrog. Energy **34**(9), 3734–3742 (2009). https://doi.org/10.1016/j.ijhydene.2009.02.074

84. J.F. Múnera et al., Combined oxidation and reforming of methane to produce pure H2 in a membrane reactor. Chem. Eng. J. **161**(1), 204–211 (2010). https://doi.org/10.1016/j.cej.2010.04.022

85. Y. Li et al., Oxidative reformings of methane to syngas with steam and CO2 catalyzed by metallic Ni based monolithic catalysts. Catal. Commun. **9**(6), 1040–1044 (2008). https://doi.org/10.1016/j.catcom.2007.10.003

86. C. Jensen, M.S. Duyar, Thermodynamic analysis of dry reforming of methane for valorization of landfill gas and natural gas. Energy Technol. **9**(7) (2021). https://doi.org/10.1002/ente.202100106

87. J. Hunt et al., Microwave-specific enhancement of the carbon–carbon dioxide (Boudouard) reaction. J. Phys. Chem. C **117**(51), 26871–26880 (2013). https://doi.org/10.1021/jp4076965

88. A.T. Bell, The impact of nanoscience on heterogeneous catalysis. Science **299**(5613), 1688–1691 (2003). https://doi.org/10.1126/science.1083671

89. I. Rivas et al., Perovskite-type oxides in methane dry reforming: effect of their incorporation into a mesoporous SBA-15 silica-host. Catal. Today **149**(3), 388–393 (2010). https://doi.org/10.1016/j.cattod.2009.05.028

90. H.R. Godini et al., Techno-economic analysis of integrating the methane oxidative coupling and methane reforming processes. Fuel Process. Technol. **106**, 684–694 (2013). https://doi.org/10.1016/j.fuproc.2012.10.002

91. K. Mondal et al., Dry reforming of methane to syngas: A potential alternative process for value added chemicals—a techno-economic perspective. Environ. Sci. Pollut. Res. **23**(22), 22267–22273 (2016). https://doi.org/10.1007/s11356-016-6310-4

92. S. Kim et al., Techno-economic analysis (tea) for CO2 reforming of methane in a membrane reactor for simultaneous CO2 utilization and ultra-pure H2 production. Int. J. Hydrog. Energy **43**(11), 5881–5893 (2018). https://doi.org/10.1016/j.ijhydene.2017.09.084

93. Q. Chen et al., Techno-economic evaluation of CO2-rich natural gas dry reforming for linear alpha olefins production. Energy Convers. Manage. **205**, 112348 (2020). https://doi.org/10.1016/j.enconman.2019.112348

94. M.M. Ramirez-Corredores et al., Identification of opportunities for integrating chemical processes for carbon (dioxide) utilization to nuclear power plants. Renew. Sustain. Energy Rev. **150**(111450), 15 (2021). https://doi.org/10.1016/j.rser.2021.111450

95. X. Gao et al., Smart designs of anti-coking and anti-sintering Ni-based catalysts for dry reforming of methane: a recent review. Reactions **1**(2), 162–194 (2020)

96. T.A. Chubb, Characteristics of CO2-CH4 reforming-methanation cycle relevant to the solchem thermochemical power system. Sol. Energy **24**(4), 341–345 (1980). https://doi.org/10.1016/0038-092X(80)90295-9

97. J.D. Fish, D.C. Hawn, Closed loop thermochemical energy transport based on CO2 reforming of methane: Balancing the reaction systems. J. Sol.Energy Eng. **109**(3), 215–220 (1987). https://doi.org/10.1115/1.3268209

98. S.C. Teuner et al., CO through CO2 reforming - the calcor standard and calcor economy processes. Oil Gas European Magazine **27**(3), 44–46 (2001)

99. G. Kurz, S. Teuner, *Calcor Process for CO Production*, vol. 43 (Erdoel und Kohle, Erdgas, Petrochemie, Germany, F.R., 1990), pp. 171–172.

100. N.R. Udengaard, Sulfur passivated reforming process lowers syngas H2/CO ratio. Oil Gas J. **90**(10), 62–67 (1992)

101. J.R. Rostrup-Nielsen, Sulfur-passivated nickel catalysts for carbon-free steam reforming of methane. J. Catal. **85**(1), 31–43 (1984). https://doi.org/10.1016/0021-9517(84)90107-6

102. H.C. Dibben et al., Make low H2/CO syngas using sulfur passivated reforming. Hydrocarbon Process **65**(1), 71–74 (1986)

103. P.M. Mortensen, I. Dybkjær, Industrial scale experience on steam reforming of CO2-rich gas. Appl. Catal. A **495**, 141–151 (2015). https://doi.org/10.1016/j.apcata.2015.02.022

104. P. Weiland, Biogas production: current state and perspectives. Appl. Microbiol. Biotechnol. **85**(4), 849–860 (2010). https://doi.org/10.1007/s00253-009-2246-7
105. F.M. Baena-Moreno et al., Analysis of the potential for biogas upgrading to syngas via catalytic reforming in the United kingdom. Renew. Sustain. Energy Rev. **144** (2021). https://doi.org/10.1016/j.rser.2021.110939
106. R.Y. Chein, W.H. Hsu, Analysis of syngas production from biogas via the tri-reforming process. Energies **11** (5) (2018). https://doi.org/10.3390/en11051075
107. A. Boretti et al., Hydrogen production by solar thermochemical water-splitting cycle via a beam down concentrator. Front. Energy Res. **9**, 5 (2021). https://doi.org/10.3389/fenrg.2021.666191
108. L.C. Buelens et al., 110th anniversary: carbon dioxide and chemical looping: current research trends. Ind. Eng. Chem. Res. **58**(36), 16235–16257 (2019). https://doi.org/10.1021/acs.iecr.9b02521
109. L.S. Fan, F. Li, Chemical looping technology and its fossil energy conversion applications. Ind. Eng. Chem. Res. **49**(21), 10200–10211 (2010). https://doi.org/10.1021/ie1005542
110. H. Fang et al., Advancements in development of chemical-looping combustion: a review. Int. J. Chem. Eng. **2009**, 710515 (2009). https://doi.org/10.1155/2009/710515
111. M. Osman, et al. Review of pressurized chemical looping processes for power generation and chemical production with integrated CO2 capture. Fuel Process. Technol. **214**(106684), 29 (2021) https://doi.org/10.1016/j.fuproc.2020.106684
112. L. Zhu et al., Comparison of carbon capture igcc with chemical-looping combustion and with calcium-looping process driven by coal for power generation. Chem. Eng. Res. Des. **104**, 110–124 (2015). https://doi.org/10.1016/j.cherd.2015.07.027
113. Z. Huang et al., Evaluation of multi-cycle performance of chemical looping dry reforming using CO2 as an oxidant with Fe-Ni bimetallic oxides. J. Energy Chem. **25**(1), 62–70 (2016). https://doi.org/10.1016/j.jechem.2015.10.008
114. V.V. Galvita et al., CeO2-modified Fe2O3 for CO2 utilization via chemical looping. Ind. Eng. Chem. Res. **52**(25), 8416–8426 (2013). https://doi.org/10.1021/ie4003574
115. M. Najera et al., Carbon capture and utilization via chemical looping dry reforming. Chem. Eng. Res. Des. **89**(9), 1533–1543 (2011). https://doi.org/10.1016/j.cherd.2010.12.017
116. R.C. Pullar et al., A review of solar thermochemical CO2 splitting using ceria-based ceramics with designed morphologies and microstructures. Front. Chem. **7**, 34 (2019). https://doi.org/10.3389/fchem.2019.00601
117. H. Webb et al., New applications of high-temperature solar energy for the production of transportable fuels and chemicals and for energy storage. ATR-78(7691–04)-1 Report (The Aerospace Corporation, Los Angeles, CA. USA, 1979), p. 180
118. K.R. Rout et al., Highly selective CO removal by sorption enhanced Boudouard reaction for hydrogen production. Catal. Sci. Technol. **9**(15), 4100–4107 (2019). https://doi.org/10.1039/c9cy00851a
119. R. Küngas, Review—electrochemical CO2 reduction for CO production: comparison of low- and high-temperature electrolysis technologies. J. Electrochem. Soc. **167**(4), 044508 (2020). https://doi.org/10.1149/1945-7111/ab7099
120. D. Higgins et al., Gas-diffusion electrodes for carbon dioxide reduction: a new paradigm. ACS Energy Lett. **4**(1), 317–324 (2019). https://doi.org/10.1021/acsenergylett.8b02035
121. D.S. Ripatti et al., Carbon monoxide gas diffusion electrolysis that produces concentrated C2 products with high single-pass conversion. Joule **3**(1), 240–256 (2019). https://doi.org/10.1016/j.joule.2018.10.007
122. M.E. Leonard et al., Investigating electrode flooding in a flowing electrolyte, gas-fed carbon dioxide electrolyzer. Chemsuschem **13**(2), 400–411 (2020). https://doi.org/10.1002/cssc.201902547
123. J.E. Obrien et al., *High-temperature electrolysis for large-scale hydrogen and syngas production from nuclear energy—summary of system simulation and economic analyses.* Int. J. Hydrog. Energy **35** (10), 4808–4819 (2010). https://doi.org/10.1016/j.ijhydene.2009.09.009

124. S.R. Foit et al., Power-to-syngas: an enabling technology for the transition of the energy system? Angew. Chem. Int. Ed. **56**(20), 5402–5411 (2017). https://doi.org/10.1002/anie.201 607552

125. S. Hernández et al., Syngas production from electrochemical reduction of CO2: current status and prospective implementation. Green Chem. **19**(10), 2326–2346 (2017). https://doi.org/10. 1039/C7GC00398F

126. A. Alcasabas et al., A comparison of different approaches to the conversion of carbon dioxide into useful products: Part I CO2 reduction by electrocatalytic, thermocatalytic and biological routes. Johns. Matthey Technol. Rev. **65**(2), 180–196 (2021). https://doi.org/10.1595/205651 321x16081175586719

127. S.H. Jensen et al., Hydrogen and synthetic fuel production from renewable energy sources. Int. J. Hydrog. Energy **32** (15 SPEC. ISS.), 3253–3257 (2007). https://doi.org/10.1016/j.ijh ydene.2007.04.042

128. M.S. Sohal et al., Degradation issues in solid oxide cells during high temperature electrolysis. J. Fuel Cell Sci. Technol. **9** (1) (2012). https://doi.org/10.1115/1.4003787

129. X. Zhang et al., Durability evaluation of reversible solid oxide cells. J. Power Sources **242**, 566–574 (2013). https://doi.org/10.1016/j.jpowsour.2013.05.134

130. J.E. Obrien et al., Long-term performance of solid oxide stacks with electrode-supported cells operating in the steam electrolysis mode, in *Proceedings of the ASME 2011 International Mechanical Engineering Congress and Exposition, IMECE* 2011, vol. 4 (Denver, CO 2011), pp. 495–503

131. E.A. Harvego et al., System evaluation and life-cycle cost analysis of a commercialscale high-temperature electrolysis hydrogen production plant, in *Proceedings of the ASME 2012 International Mechanical Engineering Congress and Exposition, IMECE*, vol. 6 (Houston, TX 2012), pp. 875–884. https://doi.org/10.1115/IMECE2012-89649

132. R. Küngas et al., Systematic lifetime testing of stacks in CO2 electrolysis. ECS Trans. **78**(1), 2895–2905 (2017)

133. J. Artz et al., Sustainable conversion of carbon dioxide: an integrated review of catalysis and life cycle assessment. Chem. Rev. **118**(2), 434–504 (2018). https://doi.org/10.1021/acs.che mrev.7b00435

134. L.A. Diaz et al., Electrochemical production of syngas from CO2 captured in switchable polarity solvents. Green Chem. **20**, 620–626 (2018). https://doi.org/10.1039/C7GC03069J

135. N. Gao et al., Intensified co-electrolysis process for syngas production from captured CO2. J. CO2 Util. **43**(101365), 22 (2020). https://doi.org/10.1016/j.jcou.2020.101365

136. G. Lee et al., Electrochemical upgrade of CO2 from amine capture solution. Nat. Energy (2020). https://doi.org/10.1038/s41560-020-00735-z

137. K.P. Kuhl et al., (Opus 12 Inc.) Reactor with advanced architecture for the electrochemical reaction of CO2, CO and other chemical compounds, Patent No. US2017321333, US2017321334 (Also published as WO2017192787, WO2017192788) (2017), 09 Nov 2017

138. T.E. Lister et al., (Battelle Energy Alliance, Llc) Methods and systems for the electrochemical reduction of carbon dioxide using switchable polarity materials, Patent No. US2020255958 (Also published as WO2019070526) (2020), 13 Aug 2020

139. R.I. Masel, (Dioxide Materials Inc) Devices for electrocatalytic conversion of carbon dioxide, Patent No. US10173169 (Also published as US2018111083), 26 Apr 2018

140. R.I. Masel, (Dioxide Materials Inc) Electrocatalytic process for carbon dioxide conversion, Patent No. US2019211463, 11 Jul 2019

141. R.I. Masel et al., (Dioxide Materials Inc) Electrochemical device for converting carbon dioxide to a reaction product, Patent No. US9481939 (Also published as US2016108530). 01 Nov 2016

142. R.I. Masel, A. Salehi-Khojin, (Dioxide Materials Inc) Electrocatalytic process for carbon dioxide conversion, Patent No. US9012345 (Also published as US2013157174) 21 Apr 2015

143. R.I. Masel, A. Salehi-Khojin, (Dioxide Materials Inc) Electrocatalytic process for carbon dioxide conversion, Patent No. US9555367 (Also published as US2015209722), 31 Jan 2017

144. R.I. Masel et al., (Dioxide Materials Inc) Electrocatalytic process for carbon dioxide conversion, Patent No. US9815021 (Also published as US2017259206, WO2017176600), 14 Sep 2017

145. A. Elmekawy et al., Technological advances in CO2 conversion electro-biorefinery: a step toward commercialization. Biores. Technol. **215**, 357–370 (2016). https://doi.org/10.1016/j.biortech.2016.03.023

146. M.-Y. Lee et al., Current achievements and the future direction of electrochemical CO2 reduction: a short review. Crit. Rev. Environ. Sci. Technol. **50**(8), 769–815 (2020). https://doi.org/10.1080/10643389.2019.1631991

147. P.G. Jessop et al., Recent advances in the homogeneous hydrogenation of carbon dioxide. Coord. Chem. Rev. **248**(21–24), 2425–2442 (2004). https://doi.org/10.1016/j.ccr.2004.05.019

148. W.M. Budzianowski, Value-added carbon management technologies for low CO2 intensive carbon-based energy vectors. Energy **41**(1), 280–297 (2012). https://doi.org/10.1016/j.energy.2012.03.008

149. M. Aresta, The carbon dioxide problem, in *An Economy Based on Carbon Dioxide and Water, Potential of Large Scale Carbon Dioxide Utilization*, ed. by M.K. Aresta, Iftekhar, S. Kawi (Springer, Switzerland AG, 2019), pp. v–xi

150. M. Aresta, *Carbon Dioxide as Chemical Feedstock* (Wiley-VCH, 2010), pp. 394. https://doi.org/10.1002/9783527629916

151. M. Aresta et al., An Economy Based on Carbon Dioxide and Water, Potential of Large Scale Carbon Dioxide Utilization, ed. by M. Aresta (Springer, Switzerland AG, 2019), p. 436. https://doi.org/10.1007/978-3-030-15868-2

152. S.G. Jadhav et al., Catalytic carbon dioxide hydrogenation to methanol: a review of recent studies. Chem. Eng. Res. Des. **92**(11), 2557–2567 (2014). https://doi.org/10.1016/j.cherd.2014.03.005

153. G.A. Olah et al., *Beyond Oil and Gas: The Methanol Economy*, 2nd edn. (Wiley-VCH, 2009), p. 334. https://doi.org/10.1002/9783527627806

154. F.S. Ramos et al., Role of dehydration catalyst acid properties on one-step dme synthesis over physical mixtures. Gas-to-liquid technology papers presented at the 12th Brazilian congress on catalysis, vol. 101 (1) (2005), pp. 39–44. https://doi.org/10.1016/j.cattod.2004.12.007

155. F. Gallucci et al., An experimental study of CO2 hydrogenation into methanol involving a zeolite membrane reactor. Chem. Eng. Process. **43**(8), 1029–1036 (2004). https://doi.org/10.1016/j.cep.2003.10.005

156. M.R. Gogate, Methanol-to-olefins process technology: current status and future prospects. Pet. Sci. Technol. **37**(5), 559–565 (2019). https://doi.org/10.1080/10916466.2018.1555589

157. G.K.S. Prakash et al., Beyond Oil and Gas: The Methanol Economy, in *Proceedings of the Electrochemistry and Climate Change—219th ECS Meeting*, vol. 35 (Montreal, QC, 2011), pp. 31–40. https://doi.org/10.1149/1.3645178

158. A. Simonov, Electrolysis breakthrough could solve the hydrogen conundrum (2019). https://phys.org/news/2019-09-electrolysis-breakthrough-hydrogen-conundrum.html. Accessed Jan 2022

159. Haldor Topsoe, Electrify methanol production for a sustainable business (2019). https://info.topsoe.com/emethanol. Accessed Dec 2021

160. C. Ampelli et al., An electrochemical reactor for the CO2 reduction in gas phase by using conductive polymer based electrocatalysts, in *10th European Symposium on Electrochemical Engineering, ESEE 2014*, vol. 41 (Special Issue), (2014), pp. 13–18. https://doi.org/10.3303/CET1441003

161. S. Schlager et al., Biocatalytic and bioelectrocatalytic approaches for the reduction of carbon dioxide using enzymes. Energ. Technol. **5**(6), 812–821 (2017). https://doi.org/10.1002/ente.201600610

162. S. Schlager et al., Electrochemical reduction of carbon dioxide to methanol by direct injection of electrons into immobilized enzymes on a modified electrode. Chemsuschem **9**(6), 631–635 (2016). https://doi.org/10.1002/cssc.201501496

163. W. Zhang et al., Progress and perspective of electrocatalytic CO2 reduction for renewable carbonaceous fuels and chemicals. Adv. Sci. **5**(1), 1700275 (2018). https://doi.org/10.1002/advs.201700275

164. Carbon Recycling International (CRI) and HS Orka George Olah CO2 to renewable methanol plant (2011). https://www.chemicals-technology.com/projects/george-olah-renewable-methanol-plant-iceland/. Accessed Aug 2011

165. H. Goehna, P. Koenig, Producing methanol from CO2. ChemTech **24**(6), 36–39 (1994)

166. M.T. Rodrigues et al., Rwgs reaction employing Ni/Mg(Al, Ni)O—the role of the O vacancies. Appl. Catal. A **543**, 98–103 (2017). https://doi.org/10.1016/j.apcata.2017.06.026

167. O.-S. Joo et al., Carbon dioxide hydrogenation to form methanol via a reverse-water-gas-shift reaction (the camere process). Ind. Eng. Chem. Res. **38**(5), 1808–1812 (1999). https://doi.org/10.1021/ie9806848

168. O.S. Joo et al., Camere process for methanol synthesis from CO2 hydrogenation, in *Studies in Surface Science and Catalysis* (Elsevier Inc, 2004), pp. 67–72

169. Carbon Recycling International (CRI), Emission to liquid (ETL) technology (2016). https://www.carbonrecycling.is/projects#project-goplant. Accessed Aug 2016

170. B. Anicic et al., Comparison between two methods of methanol production from carbon dioxide. Energy **77**, 279–289 (2014). https://doi.org/10.1016/j.energy.2014.09.069

171. O.Y. Abdelaziz et al, Novel process technologies for conversion of carbon dioxide from industrial flue gas streams into methanol. J. CO2 Util. **21**, 52–63 (2017). https://doi.org/10.1016/j.jcou.2017.06.018

172. D. Milani et al., A model-based analysis of CO2 utilization in methanol synthesis plant. J. CO2 Util. **10**, 12–22 (2015). https://doi.org/10.1016/j.jcou.2015.02.003

173. R.I. Masel et al., (Dioxide Materials Inc) Hydrogenation of formic acid to formaldehyde, Patent No. US9193593 (Also published as AU2014218628, CN105339336, CN107557086, EP2958882, US2014239231, WO2014130962), 24 Nov 2015

174. R.I. Masel et al., (Dioxide Materials Inc) *Devices and processes for carbon dioxide conversion into useful fuels and chemicals* Patent No. US9181625 (Also published as CN104822861, EP2898120, US2014093799, WO2014047661, WO201404766), 10 Nov 2015

175. R.I. Masel, (Dioxide Materials Inc) Process for the sustainable production of acrylic acid, Patent No. US9790161 (Also published as US2019135726 (05/09/2019), US2018057439 (03/01/18), US20160207866 (07/21/16), AU2017200539), 17 Oct 2017

176. R.I. Masel, (Dioxide Materials Inc) System and process for the production of renewable fuels and chemicals, Patent No. EP3504359 (Also published as CN109642332, KR20190043156, WO2018044720), 3 July 2019

177. J.J. Kaczur et al., (Dioxide Materials Inc) Method and system for electrochemical production of formic acid from carbon dioxide, Patent No. US2019010620, 01 Oct 2019

178. C. Toro, E. Sciubba, Sabatier based power-to-gas system: heat exchange network design and thermoeconomic analysis. Appl. Energy **229**, 1181–1190 (2018). https://doi.org/10.1016/j.apenergy.2018.08.036

179. M. Gruber et al., Integrated high-temperature electrolysis and methanation (helmeth) technology: the methanation process (2013). http://www.helmeth.eu/index.php/technologies/methanation-process. Accessed Mar 2022

180. M. Gruber et al., Power-to-gas through thermal integration of high-temperature steam electrolysis and carbon dioxide methanation—experimental results. Fuel Process. Technol. **181**, 61–74 (2018). https://doi.org/10.1016/j.fuproc.2018.09.003

181. M. Seemann, H. Thunman, *Methane Synthesis*, in *Substitute Natural Gas from Waste*, ed. by M. Materazzi, P.U. Foscolo (Academic Press, 2019), pp. 221–243. https://doi.org/10.1016/B978-0-12-815554-7.00009-X

182. D. Beierlein et al., Experimental approach for identifying hotspots in lab-scale fixed-bed reactors exemplified by the sabatier reaction. React. Kinet. Mech. Catal. **125**(1), 157–170 (2018). https://doi.org/10.1007/s11144-018-1402-4

183. E. Moioli et al., Model based determination of the optimal reactor concept for sabatier reaction in small-scale applications over Ru/Al2O3. Chem. Eng. J. 121954, 10 (2019). https://doi.org/10.1016/j.cej.2019.121954

184. E. Moioli et al., Parametric sensitivity in the sabatier reaction over Ru/Al2O3-theoretical determination of the minimal requirements for reactor activation. React. Chem. Eng. **4**(1), 100–111 (2019). https://doi.org/10.1039/c8re00133b

185. W. Wang et al., CO2 methanation under dynamic operational mode using nickel nanoparticles decorated carbon felt (Ni/OCF) combined with inductive heating. Catal. Today (2019). https://doi.org/10.1016/j.cattod.2019.02.050

186. J. Díez-Ramírez et al., Effect of support nature on the cobalt-catalyzed CO2 hydrogenation. J. CO2 Util. **21**, 562–571 (2017). https://doi.org/10.1016/j.jcou.2017.08.019

187. L. Falbo et al., Kinetics of CO2 methanation on a Ru-based catalyst at process conditions relevant for power-to-gas applications. Appl. Catal. B **225**, 354–363 (2018). https://doi.org/10.1016/j.apcatb.2017.11.066

188. P.A. Aldana et al., Catalytic CO2 valorization into CH4 on Ni-based ceria-zirconia. Reaction mechanism by operando IR spectroscopy. Catal. Today **215**, 201–207 (2013). https://doi.org/10.1016/j.cattod.2013.02.019

189. I. García–García et al., Power-to-gas: Storing surplus electrical energy. Study of Al2O3 support modification. Int. J. Hydrog. Energy **41**(43), 19587–19594 (2016). https://doi.org/10.1016/j.ijhydene.2016.04.010

190. S. Navarro-Jaén et al., Size-tailored Ru nanoparticles deposited over γ-Al2O3 for the CO2 methanation reaction. Appl. Surf. Sci. **483**, 750–761 (2019). https://doi.org/10.1016/j.apsusc.2019.03.248

191. A. Porta et al., Synthesis of Ru-based catalysts for CO2 methanation and experimental assessment of intraporous transport limitations. Catal. Today **343**, 38–47 (2020). https://doi.org/10.1016/j.cattod.2019.01.042

192. J.V. Veselovskaya et al., Catalytic methanation of carbon dioxide captured from ambient air. Energy **159**, 766–773 (2018). https://doi.org/10.1016/j.energy.2018.06.180

193. J. Xu et al., Influence of pretreatment temperature on catalytic performance of rutile TiO2-supported ruthenium catalyst in CO2 methanation. J. Catal. **333**, 227–237 (2016). https://doi.org/10.1016/j.jcat.2015.10.025

194. V. Barbarossa et al., CO2 conversion to CH4. Green Energy Technol. **137**, 123–145 (2013). https://doi.org/10.1007/978-1-4471-5119-7_8

195. T. Schaaf et al., Methanation of CO2 - storage of renewable energy in a gas distribution system. Energy Sustain. Soc. **4**(1) (2014). https://doi.org/10.1186/s13705-014-0029-1

196. A. Lazdans et al., Development of the experimental scheme for methanation process. Energy Procedia **95**, 540–545 (2016). https://doi.org/10.1016/j.egypro.2016.09.082

197. D. Schlereth, O. Hinrichsen, A fixed-bed reactor modeling study on the methanation of CO2. Chem. Eng. Res. Des. **92**(4), 702–712 (2014). https://doi.org/10.1016/j.cherd.2013.11.014

198. S.K. Hoekman et al., CO2 recycling by reaction with renewably-generated hydrogen. Int. J. Greenh. Gas Control **4**(1), 44–50 (2010). https://doi.org/10.1016/j.ijggc.2009.09.012

199. J.V. Veselovskaya et al., A novel process for renewable methane production: combining direct air capture by K2CO3/alumina sorbent with CO2 methanation over Ru/alumina catalyst. Top. Catal. **61**(15–17), 1528–1536 (2018). https://doi.org/10.1007/s11244-018-0997-z

200. S. Walspurger et al., Sorption enhanced methanation for substitute natural gas production: experimental results and thermodynamic considerations. Chem. Eng. J. **242**, 379–386 (2014). https://doi.org/10.1016/j.cej.2013.12.045

201. D. Schollenberger et al., Scale-up of innovative honeycomb reactors for power-to-gas applications—the project store&go. Chem.-Ing.-Tech. **90**(5), 696–702 (2018). https://doi.org/10.1002/cite.201700139

202. C. Bassano et al., P2G movable modular plant operation on synthetic methane production from CO2 and hydrogen from renewables sources. Fuel **253**, 1071–1079 (2019). https://doi.org/10.1016/j.fuel.2019.05.074

203. K. Stangeland et al., CO2 methanation: the effect of catalysts and reaction conditions. Energy Procedia **105**, 2022–2027 (2017). https://doi.org/10.1016/j.egypro.2017.03.577

204. I. García-García et al., Power-to-gas: Storing surplus electrical energy. Study of catalyst synthesis and operating conditions. Int. J. Hydrog. Energy **43**(37), 17737–17747 (2018). https://doi.org/10.1016/j.ijhydene.2018.06.192

205. J. Uebbing et al., Exergetic assessment of CO2 methanation processes for the chemical storage of renewable energies. Appl. Energy **233–234**, 271–282 (2019). https://doi.org/10.1016/j.apenergy.2018.10.014

206. M. Bailera et al., Future applications of hydrogen production and CO2 utilization for energy storage: hybrid power to gas-oxycombustion power plants. Int. J. Hydrog Energy **42**(19), 13625–13632 (2017). https://doi.org/10.1016/j.ijhydene.2017.02.123

207. M. De Saint Jean et al., Parametric study of an efficient renewable power-to-substitute-natural-gas process including high-temperature steam electrolysis. Int. J. Hydrog Energy **39**(30), 17024–17039 (2014). https://doi.org/10.1016/j.ijhydene.2014.08.091

208. J. Bremer et al., CO2 methanation: Optimal start-up control of a fixed-bed reactor for power-to-gas applications. AIChE J. **63**(1), 23–31 (2017). https://doi.org/10.1002/aic.15496

209. G. Leonzio, Design and feasibility analysis of a power-to-gas plant in germany. J. Clean. Prod. **162**, 609–623 (2017). https://doi.org/10.1016/j.jclepro.2017.05.168

210. L. Jürgensen et al., Utilization of surplus electricity from wind power for dynamic biogas upgrading: Northern germany case study. Biomass Bioenerg. **66**, 126–132 (2014). https://doi.org/10.1016/j.biombioe.2014.02.032

211. L. Jürgensen et al., Dynamic biogas upgrading based on the sabatier process: thermodynamic and dynamic process simulation. Biores. Technol. **178**, 323–329 (2015). https://doi.org/10.1016/j.biortech.2014.10.069

212. L. Jürgensen et al., Influence of trace substances on methanation catalysts used in dynamic biogas upgrading. Biores. Technol. **178**, 319–322 (2015). https://doi.org/10.1016/j.biortech.2014.09.080

213. B. Castellani et al., Carbon and energy footprint of the hydrate-based biogas upgrading process integrated with CO2 valorization. Sci. Total Environ. **615**, 404–411 (2018). https://doi.org/10.1016/j.scitotenv.2017.09.254

214. T.T.Q. Vo et al., Techno-economic analysis of biogas upgrading via amine scrubber, carbon capture and ex situ methanation. Appl. Energy **212**, 1191–1202 (2018). https://doi.org/10.1016/j.apenergy.2017.12.099

215. S. Falcinelli, Fuel production from waste CO2 using renewable energies. Catal. Today **348**, 95–101 (2020). https://doi.org/10.1016/j.cattod.2019.08.041

216. G. Borrel et al., Methanogenesis and the wood–ljungdahl pathway: an ancient, versatile, and fragile association. Genome Biol. Evol. **8**(6), 1706–1711 (2016). https://doi.org/10.1093/gbe/evw114

217. P.J. Strong et al., A methanotroph-based biorefinery: Potential scenarios for generating multiple products from a single fermentation. Biores. Technol. **215**, 314–323 (2016). https://doi.org/10.1016/j.biortech.2016.04.099

218. S. Nariya, M.G. Kalyuzhnaya, *Diversity, Physiology, and Biotechnological Potential of halo(alkali)philic Methane-Consuming Bacteria, in Methanotrophs: Microbiology Fundamentals and Biotechnological Applications,* ed. by E.Y. Lee (Springer International Publishing, Cham, 2019), pp. 139–161. https://doi.org/10.1007/978-3-030-23261-0_5

219. M. Van Dael et al., Techno-economic assessment of a microbial power-to-gas plant—case study in belgium. Appl. Energy **215**, 416–425 (2018). https://doi.org/10.1016/j.apenergy.2018.01.092

220. E. Giglio et al., Synthetic natural gas via integrated high-temperature electrolysis and methanation: Part I-energy performance. J. Energy Storage **1**(1), 22–37 (2015). https://doi.org/10.1016/j.est.2015.04.002

221. D. Sun, D.S.A. Simakov, Thermal management of a sabatier reactor for CO2 conversion into CH4: Simulation-based analysis. J. CO2 Util. **21**, 368–382 (2017). https://doi.org/10.1016/j.jcou.2017.07.015

222. M. Thema et al., Power-to-gas: electrolysis and methanation status review. Renew. Sustain. Energy Rev. **112**, 775–787 (2019). https://doi.org/10.1016/j.rser.2019.06.030

223. M.V. Solmi et al. CO2 as a building block for the catalytic synthesis of carboxylic acids. in *Studies in Surface Science and Catalysis.* (Elsevier Inc, 2019), pp. 105–124. https://doi.org/10.1016/B978-0-444-64127-4.00006-9

224. M. Hölscher et al., Carbon dioxide as a carbon resource—recent trends and perspectives. Z. Fur Naturforschung—Section B J. Chem. Sci. **67**(10), 961–975 (2012). https://doi.org/10. 5560/ZNB.2012-0219

225. M. Scott et al., Methylformate from CO2: an integrated process combining catalytic hydrogenation and reactive distillation. Green Chem. **21**(23), 6307–6317 (2019). https://doi.org/ 10.1039/c9gc03006a

226. M. Schmitz et al., Catalytic processes combining CO2 and alkenes into value-added chemicals: synthesis of cyclic carbonates, lactones, carboxylic acids, esters, aldehydes, alcohols, and amines. Top. Organomet. Chem. **63**, 17–38 (2019). https://doi.org/10.1007/3418_2018_24

227. M. Peters et al., Chemical technologies for exploiting and recycling carbon dioxide into the value chain. Chemsuschem **4**(9), 1216–1240 (2011). https://doi.org/10.1002/cssc.201000447

228. T.G. Ostapowicz et al., Carbon dioxide as a C1 building block for the formation of carboxylic acids by formal catalytic hydrocarboxylation. Angew. Chem. Int. Ed. **52**(46), 12119–12123 (2013). https://doi.org/10.1002/anie.201304529

229. J. Klankermayer, W. Leitner, Love at second sight for CO2 and H2 in organic synthesis. Science **350**(6261), 629–630 (2015). https://doi.org/10.1126/science.aac7997

230. J. Kizlink, I. Pastucha, Preparation of dimethyl carbonate from methanol and carbon dioxide in the presence of Sn(IV) and ti(IV) alkoxides and metal acetates. Collect. Czech. Chem. Commun. **60**, 687–692 (1995)

231. D. Ballivet-Tkatchenko et al., Carbon dioxide conversion to dimethyl carbonate: the effect of silica as support for SnO2 and ZrO2 catalysts. C. R. Chim. **14**(7), 780–785 (2011). https:// doi.org/10.1016/j.crci.2010.08.003

232. D. Delledonne et al., Developments in the production and application of dimethylcarbonate. Appl. Catal. A **221**(1), 241–251 (2001). https://doi.org/10.1016/S0926-860X(01)00796-7

233. P. Tundo et al., Synthesis of carbamates from amines and dialkyl carbonates: influence of leaving and entering groups. Synlett **10**, 1567–1571 (2010). https://doi.org/10.1055/s-0029-1219927

234. P. Tundo et al., *The greening of chemistry*, in *The chemical element: Chemistry's contribution to our global future*, 1st edn. (Wiley-VCH. 2011), pp. 189–233. https://doi.org/10.1002/978 3527635641.ch6

235. E. Alper, O. Yuksel Orhan, CO2 utilization: developments in conversion processes. Petroleum **3**(1), 109–126 (2017). https://doi.org/10.1016/j.petlm.2016.11.003

236. S. Liang et al., Highly efficient synthesis of cyclic carbonates from CO2 and epoxides over cellulose/ki. Chem. Commun. **47**(7), 2131–2133 (2011). https://doi.org/10.1039/c0cc04829a

237. S. Elmas et al., Anion effect controlling the selectivity in the zinc-catalysed copolymerisation of CO2 and cyclohexene oxide. Beilstein J. Org. Chem. **11**, 42–49 (2015). https://doi.org/10. 3762/bjoc.11.7

238. J. Langanke et al., Polymers from CO2—an industrial perspective, in *Carbon Dioxide Utilisation*, ed. by P. Styring et al. (Elsevier, Amsterdam, 2015), pp. 59–71. https://doi.org/10. 1016/B978-0-444-62746-9.00005-0

239. W. Leitner et al., How can we put the climate killer CO2 to good use? Bayer Annual Report (2010), pp. 32–37.

240. W. Leitner et al., Carbon2polymer—chemical utilization of CO2 in the production of isocyanates. Chem.-Ing.-Tech. **90**(10), 1504–1512 (2018). https://doi.org/10.1002/cite.201 800040

241. H.S. Suh et al., Polyester polyol synthesis by alternating copolymerization of propylene oxide with cyclic acid anhydrides by using double metal cyanide catalyst. React. Funct. Polym. **70**(5), 288–293 (2010). https://doi.org/10.1016/j.reactfunctpolym.2010.02.001

242. M. Aresta, Perspectives of carbon dioxide utilisation in the synthesis of chemicals, coupling chemistry with biotechnology. Stud. Surf. Sci. Catal. **114**, 65–76 (1998). https://doi.org/10. 1016/s0167-2991(98)80727-1

243. M. Aresta et al., State of the art and perspectives in catalytic processes for CO2 conversion into chemicals and fuels: the distinctive contribution of chemical catalysis and biotechnology. J. Catal. **343**, 2–45 (2016). https://doi.org/10.1016/j.jcat.2016.04.003

244. M. Aresta et al., Biotechnology to develop innovative syntheses using CO2. Environ. Chem. Lett. **3**(3), 113–117 (2005). https://doi.org/10.1007/s10311-005-0009-y

245. W.W. Cleland et al., Mechanism of rubisco: the carbamate as general base. Chem. Rev. **98**(2), 549–561 (1998). https://doi.org/10.1021/cr970010r

246. N. Getoff, G.O. Schenck, On the formation of amino acids by gamma-ray-induced carboxylation of amines in aqueous solutions. Radiat. Res. **31**(3), 486–505 (1967). https://doi.org/10.2307/3572367

247. N. Getoff, Carboxylation of formic acid in aqueous solutions under the influence of UV-light. Photochem. Photobiol. **4**(3), 433–438 (1965). https://doi.org/10.1111/j.1751-1097.1965.tb09756.x

248. F. Gütlbauer, N. Getoff, Radiation chemical carboxylation of hydroxycompounds. Int. J. Appl. Radiat. Isot. **16**(12), 673–680 (1965)

249. K.F. Krapfenbauer, N. Getoff, Radiation-and photo-induced formation of salicylic acid from phenol and CO2 in aqueous solution. World Resour. Rev. **9**, 421–433 (1997)

250. N. Getoff, Radiation chemistry and the environment. Radiat. Phys. Chem. **54**(4), 377–384 (1999). https://doi.org/10.1016/S0969-806X(98)00266-7

251. N. Getoff et al., One-electron oxidation of mitomycin C and its corresponding peroxyl radicals. A steady-state and pulse radiolysis study. Radiat. Phys. Chem. **50**(6), 575–583 (1997). https://doi.org/10.1016/S0969-806X(97)00099-6

252. N. Getoff et al., Pulse radiolysis of pyrene in aprotic polar organic solvents: simultaneous formation of pyrene radical cations and radical anions. Radiat. Phys. Chem. **66**(3), 207–214 (2003). https://doi.org/10.1016/S0969-806X(02)00392-4

253. M. Aresta et al., *Biorefinery: From Biomass to Chemicals and Fuels* (De Gruyter, 2012), p. 464. https://doi.org/10.1515/9783110260281

254. K. Özdenkçi et al., A novel biorefinery integration concept for lignocellulosic biomass. Energy Convers. Manage. **149**, 974–987 (2017). https://doi.org/10.1016/j.enconman.2017.04.034

255. A.J. Ragauskas, All about biorefining, in *Proceedings of the 239th ACS National Meeting* (San Francisco, CA 2010)

256. M. Valdivia et al., Biofuels 2020: biorefineries based on lignocellulosic materials. Microb. Biotechnol. **9**(5), 585–594 (2016). https://doi.org/10.1111/1751-7915.12387

257. A. Demirbas, Biorefineries: current activities and future developments. Energy Convers. Manage. **50**(11), 2782–2801 (2009)

258. E. De Jong, G. Jungmeier, Biorefinery concepts in comparison to petrochemical refineries, in *Industrial Biorefineries and White Biotechnology*, ed. by R. Höfer et al. (Elsevier, Amsterdam, The Netherlands, 2015), pp. 3–33. https://doi.org/10.1016/B978-0-444-63453-5.00001-X

259. E.F. Sousa-Aguiar et al., Petrobras: the concept of integrated biorefineries applied to the oleochemistry industry: rational utilization of products and residues via catalytic routes, in *Industrial Biorenewables: A Practical Viewpoint*, ed. by P. Domínguez De María (Wiley, 2016), pp. 451–465. https://doi.org/10.1002/9781118843796.ch20

260. E.F. Sousa-Aguiar et al., Chapter 1: Catalysts for Co-processing biomass in oil refining industry, in *RSC Green Chemistry*, ed. by F. Frusteri et al. (Royal Society of Chemistry, 2018), pp. 1–24. https://doi.org/10.1039/9781788013567-00001

261. L.P. Ozorio et al., Metal-impregnated zeolite y as efficient catalyst for the direct carbonation of glycerol with CO2. Appl. Catal. A **504**, 187–191 (2015). https://doi.org/10.1016/j.apcata.2014.12.010

262. M. North et al., Synthesis of cyclic carbonates from epoxides and CO2. Green Chem. **12**(9), 1514–1539 (2010). https://doi.org/10.1039/c0gc00065e

263. L.N. He et al., New procedure for recycling homogeneous catalyst: propylene carbonate synthesis under supercritical CO2 conditions. Green Chem. **5**(1), 92–94 (2003). https://doi.org/10.1039/b210007j

264. S. Fukuoka et al., A novel non-phosgene polycarbonate production process using by-product CO2 as starting material. Green Chem. **5**(5), 497–507 (2003). https://doi.org/10.1039/b304963a

265. T. Sakakura et al., Synthesis of dimethyl carbonate from carbon dioxide: catalysis and mechanism. Polyhedron **19**(5), 573–576 (2000). https://doi.org/10.1016/S0277-5387(99)00411-8

266. M.A. Pacheco, C.L. Marshall, Review of dimethyl carbonate (dmc) manufacture and its characteristics as a fuel additive. Energy Fuels **11**(1), 2–29 (1997). https://doi.org/10.1021/ef9600974

267. C.H. Zhou et al., Chemoselective catalytic conversion of glycerol as a biorenewable source to valuable commodity chemicals. Chem. Soc. Rev. **37**(3), 527–549 (2008). https://doi.org/10.1039/b707343g

268. M. Pagliaro, M. Rossi, *The Future of Glycerol: New Usages for a Versatile Raw Material* (RSC Publishing, Cambridge, UK, 2008), pp. 127. https://doi.org/10.1039/9781847558305

269. Y. Zheng et al., Commodity chemicals derived from glycerol, an important biorefinery feedstock. Chem. Rev. **110**(3), 1807–1807 (2010). https://doi.org/10.1021/cr068216s

270. S. Christy et al., Recent progress in the synthesis and applications of glycerol carbonate. Curr. Opin. Green Sustain. Chem. **14**, 99–107 (2018). https://doi.org/10.1016/j.cogsc.2018.09.003

271. J.W. Yoo, Z. Mouloungui, Catalytic carbonylation of glycerin by urea in the presence of zinc mesoporous system for the synthesis of glycerol carbonate, in *Studies in Surface Science and Catalysis* (Elsevier Inc, 2003), pp. 757–760. https://doi.org/10.1016/S0167-2991(03)80494-9

272. L. Wang et al., Efficient synthesis of glycerol carbonate from glycerol and urea with lanthanum oxide as a solid base catalyst. Catal. Commun. **12**(15), 1458–1462 (2011). https://doi.org/10.1016/j.catcom.2011.05.027

273. C. Hammond et al., Synthesis of glycerol carbonate from glycerol and urea with gold-based catalysts. Dalton Trans. **40**(15), 3927–3937 (2011). https://doi.org/10.1039/c0dt01389g

274. A. Dibenedetto et al., Direct carboxylation of alcohols to organic carbonates: comparison of the group 5 element alkoxides catalytic activity. An insight into the reaction mechanism and its key steps. Catal. Today **115**(1–4), 88–94 (2006). https://doi.org/10.1016/j.cattod.2006.02.026

275. M. Aresta et al., A study on the carboxylation of glycerol to glycerol carbonate with carbon dioxide: the role of the catalyst, solvent and reaction conditions. J. Mol. Catal. A: Chem. **257**(1–2), 149–153 (2006). https://doi.org/10.1016/j.molcata.2006.05.021

276. C. Vieville et al., Synthesis of glycerol carbonate by direct carbonatation of glycerol in supercritical CO2 in the presence of zeolites and ion exchange resins. Catal. Lett. **56**(4), 245–247 (1998). https://doi.org/10.1023/A:1019050205502

277. M.M. Ramirez-Corredores et al., Radiation-induced chemistry of carbon dioxide: a pathway to close the carbon loop for a circular economy. Front. Energy Res. **8**(108), 17 (2020). https://doi.org/10.3389/fenrg.2020.00108

278. A.M.A. El Naggar et al., Biomass to fuel gas conversion through a low pyrolysis temperature induced by gamma radiation: an experimental and simulative study. RSC Adv. **5**(95), 77897–77905 (2015). https://doi.org/10.1039/c5ra10621d

279. J. Cheng et al., Enhancing growth rate and lipid yield of chlorella with nuclear irradiation under high salt and CO2 stress. Biores. Technol. **203**, 220–227 (2016). https://doi.org/10.1016/j.biortech.2015.12.032

280. T. Nagai, N. Suzuki, The radiation-induced degradation of lignin in aqueous solutions. Int. J. Appl. Radiat. Isot. **29**(4–5), 255–259 (1978). https://doi.org/10.1016/0020-708X(78)90051-0

281. S.J. Zhang et al., Degradation of calcium lignosulfonate using gamma-ray irradiation. Chemosphere **57**(9), 1181–1187 (2004). https://doi.org/10.1016/j.chemosphere.2004.08.015

282. A.V. Ponomarev, B.G. Ershov, Radiation-induced high-temperature conversion of cellulose. Molecules **19**(10), 16877–16908 (2014). https://doi.org/10.3390/molecules191016877

283. C. Wang et al., Bioinspired artificial photosynthetic systems. Tetrahedron **76**(14), 131024, 15 (2022). https://doi.org/10.1002/chem.202103595

284. S. Kim et al., Artificial photosynthesis for high-value-added chemicals: old material, new opportunity. Carbon Energy **4**(1), 21–44 (2022). https://doi.org/10.1002/cey2.159

285. C.I. Ezugwu et al., Engineering metal-organic frameworks for efficient photocatalytic conversion of CO2 into solar fuels. Coord. Chem. Rev. **450**, 214245, 24 (2022). https://doi.org/10.1016/j.ccr.2021.214245

286. P. Singh, R. Srivastava, Utilization of bio-inspired catalyst for CO2 reduction into green fuels: recent advancement and future perspectives. J. CO2 Util. **53**, 101748, 25 (2021). https://doi.org/10.1016/j.jcou.2021.101748

287. N.S. Weliwatte, S.D. Minteer, Photo-bioelectrocatalytic CO2 reduction for a circular energy landscape. Joule **5**(10), 2564–2592 (2021). https://doi.org/10.1016/j.joule.2021.08.003

288. J. Liu et al., Hybrid artificial photosynthetic systems constructed using quantum dots and molecular catalysts for solar fuel production: development and advances. J. Mater. Chem. A **9**(35), 19346–19368 (2021). https://doi.org/10.1039/d1ta02673a

289. H. Bian et al., Perovskite—a wonder catalyst for solar hydrogen production. J. Energy Chem. **57**, 325–340 (2021). https://doi.org/10.1016/j.jechem.2020.08.057

290. M. Jia et al., Metal-organic frameworks and their derivatives-modified photoelectrodes for photoelectrochemical applications. Coord. Chem. Rev. **434**, 213780, 19 (2021). https://doi.org/10.1016/j.ccr.2021.213780

291. I. Bernhardsgrütter et al., CO2-converting enzymes for sustainable biotechnology: from mechanisms to application. Curr. Opin. Biotechnol. **67**, 80–87 (2021). https://doi.org/10.1016/j.copbio.2021.01.003

292. T. Keijer et al., Supramolecular strategies in artificial photosynthesis. Chem. Sci. **12**(1), 50–70 (2021). https://doi.org/10.1039/d0sc03715j

293. Z. Zhang et al., Artificial photosynthesis over metal halide perovskites: achievements, challenges, and prospects. J. Phys. Chem. Lett. **12**, 5864–5870 (2021). https://doi.org/10.1021/acs.jpclett.1c01527

294. K. Kamiya et al., CO2 electrolysis in integrated artificial photosynthesis systems. Chem. Lett. **50**(1), 166–179 (2021). https://doi.org/10.1246/CL.200691

295. R.S. Sherbo et al. Hybrid biological-inorganic systems for CO2 conversion to fuels. RSC Energy Environ. Ser. 317–346 (2021). https://doi.org/10.1039/9781788015844-00317

296. K. Amulya, S. Venkata Mohan, Augmenting succinic acid production by bioelectrochemical synthesis: influence of applied potential and CO2 availability. Chem. Eng. J. **411**, 128377, 10 (2021). https://doi.org/10.1016/j.cej.2020.128377

297. A. Alcasabas et al., A comparison of different approaches to the conversion of carbon dioxide into useful products: Part II more routes to CO2 reduction. Johnson Matthey Technol. Rev. **65**(2), 197–206 (2021). https://doi.org/10.1595/205651321x16112390198879

298. M. Levy et al., Solar energy storage via a closed-loop chemical heat pipe. Sol. Energy **50**(2), 179–189 (1993). https://doi.org/10.1016/0038-092X(93)90089-7

299. K.B. Naceur et al., Energy technology perspectives 2017: Catalysing energy technology transformations (International Energy Agency, IEA, Energy Technology Policy Division. Paris, France, 2017), p. 440. https://www.iea.org/reports/energy-technology-perspectives-2017

300. O. Edenhofer et al., *Climate Change 2014: Mitigation of Climate Change* (Intergovernmental Panel on Climate Change (IPCC), Cambridge, United Kingdom and New York, NY, USA, 2014), p. 1454. https://www.ipcc.ch/report/ar5/wg3/

301. T.B. Johansson et al., *Global Energy Assessment—Toward a Sustainable Future* (Cambridge University Press, Cambridge, UK and New York, NY, USA, 2012), p. 1884

302. A.D. Pee et al., *Decarbonization of Industrial Sectors: The Next Frontier*. (McKinsey Co, USA, 2018), p. 68

303. S.J.G. Cooper, G.P. Hammond, Decarbonising UK industry: towards a cleaner economy. Proc. Inst. Civ. Eng.: Energy **171**(4), 147–157 (2018). https://doi.org/10.1680/jener.18.00007

304. M. Andrei et al., Decarbonization of industry: Guidelines towards a harmonized energy efficiency policy program impact evaluation methodology. Energy Rep. **7**, 1385–1395 (2021). https://doi.org/10.1016/j.egyr.2021.02.067

305. E. Andersson et al., Decarbonization of industry: implementation of energy performance indicators for successful energy management practices in kraft pulp mills. Energy Rep. **7**, 1808–1817 (2021). https://doi.org/10.1016/j.egyr.2021.03.009

306. J. Xiao et al., Decarbonizing China's power sector by 2030 with consideration of technological progress and cross-regional power transmission. Energy Policy **150** (2021). https://doi.org/10.1016/j.enpol.2021.112150

307. E. Kriegler, et al. Pathways limiting warming to 1.5 °C: a tale of turning around in no time? Philos. Trans. R. Soc. A: Math., Phys. Eng. Sci. **376**(2119), 20160457 (2018). https://doi.org/ 10.1098/rsta.2016.0457

308. M. Krishnan et al., *The net-zero transition: what it would cost, what it could bring* (McKinsey Global Institute, 2022), p. 224

309. Z. Fan, S.J. Friedmann, Low-carbon production of iron and steel: technology options, economic assessment, and policy. Joule **5**(4), 829–862 (2021). https://doi.org/10.1016/j.joule. 2021.02.018

310. M. Axelson et al., Emission reduction strategies in the eu steel industry: implications for business model innovation. J. Ind. Ecol. **25**(2), 390–402 (2021). https://doi.org/10.1111/jiec. 13124

311. N. Fackler et al., Stepping on the gas to a circular economy: accelerating development of carbon-negative chemical production from gas fermentation. Annu. Rev. Chem. Biomol. Eng. **12**, 439–470 (2021). https://doi.org/10.1146/annurev-chembioeng-120120-021122

312. M. Köpke, S.D. Simpson, Pollution to products: recycling of 'above ground' carbon by gas fermentation. Curr. Opin. Biotechnol. **65**, 180–189 (2020). https://doi.org/10.1016/j.copbio. 2020.02.017

313. J. Daniell et al., Low-carbon fuel and chemical production by anaerobic gas fermentation, in *Advances in Biochemical Engineering/Biotechnology* (Springer Science and Business Media Deutschland GmbH, 2016), pp. 293–321. https://doi.org/10.1007/10_2015_5005

314. S.D. Simpson et al., (Lanzatech New Zealand Ltd) Carbon capture in fermentation, Patent No. US8376736 (Also published as US20100317074), 19 Feb 2013

315. S.D. Simpson et al., (Lanzatech New Zealand Ltd) Microbial fermentation of gaseous substrates to produce alcohols, Patent No. US7972824 (Also published as NZ546496, US2009203100, WO2007117157). 05 Jul 201.

316. S.D. Simpson et al., (Lanzatech New Zealand Ltd) Alcohol production process, Patent No. US8293509 (Also published as NZ553984, US20100105115). 23 Oct 2012.

317. S.D. Simpson et al., (Lanzatech New Zealand Ltd) Bacteria and methods of use thereof, Patent No. US8222013. 17 Jul 2012.

318. S.D. Simpson et al., (Lanzatech New Zealand Ltd) Process for production of alcohols by microbial fermentation, Patent No. US8658408. 25 Feb 2014.

319. F. Liew et al., Gas fermentation-a flexible platform for commercial scale production of low-carbon-fuels and chemicals from waste and renewable feedstocks. Front. Microbiol. **7**, 28 (2016). https://doi.org/10.3389/fmicb.2016.00694

320. NOAA, Trends in atmospheric carbon dioxide (2022). https://gml.noaa.gov/ccgg/trends/. Accessed Feb 2022

321. P.S. Adam et al., Evolutionary history of carbon monoxide dehydrogenase/acetyl-coa synthase, one of the oldest enzymatic complexes. Proc. Natl. Acad. Sci. **115**(6), E1166–E1173 (2018). https://doi.org/10.1073/pnas.1716667115

322. A.G. Fast, E.T. Papoutsakis, Stoichiometric and energetic analyses of non-photosynthetic CO2-fixation pathways to support synthetic biology strategies for production of fuels and chemicals. Curr. Opin. Chem. Eng. **1**(4), 380–395 (2012). https://doi.org/10.1016/j.coche. 2012.07.005

323. S.W. Ragsdale, E. Pierce, Acetogenesis and the wood–ljungdahl pathway of CO2 fixation. Biochim. Biophys. Acta (BBA)—Proteins Proteomics. **1784**(12), 1873–1898 (2008). https:// doi.org/10.1016/j.bbapap.2008.08.012

324. M. Can et al., Structure, function, and mechanism of the nickel metalloenzymes, CO dehydrogenase, and Acetyl-CoA synthase. Chem. Rev. **114**(8), 4149–4174 (2014). https://doi.org/ 10.1021/cr400461p

325. K. Valgepea, et al. H2 drives metabolic rearrangements in gas-fermenting clostridium autoethanogenum. Biotechnol. Biofuels **11**(1), 55, 15 (2018). https://doi.org/10.1186/s13068-018-1052-9

326. J. Desai et al., Carbon dioxide sequestration by mineral carbonation using alkaline rich material, in *Proceedings of the 2020 International Conference on Sustainable Innovations in Civil*

and Mechanical Engineering, ICSICME 2020, vol. 814. (Institute of Physics Publishing, 2020). https://doi.org/10.1088/1757-899X/814/1/012035

327. F.T.C. Röben et al., Decarbonizing copper production by power-to-hydrogen: a techno-economic analysis. J. Clean. Prod. **306**(2021). https://doi.org/10.1016/j.jclepro.2021.127191

328. R. Zevenhoven et al., Mineralization of CO2 using serpentinite rock—towards industrial application, in *Natural Resources: Sustainable Targets, Technologies, Lifestyles and Governance*, ed. by C. Ludwig et al. (World Resources Forum, 2015), pp. 125–129. http://infoscience.epfl.ch/record/213010/files/Ludwig_Natural_Resources_2015.pdf

329. A. Fernandez, K. West, *Technology Roadmap—Low-Carbon Transition in the Cement Industry.* (International Energy Agency, Paris, France, 2018), p. 62.

330. Research Institute of Innovative Technology for the Earth, Substantial GHG emissions reduction in the cement industry by using waste heat recovery combined with mineral carbon capture and utilization. Climate Technology Center Network: Southafrica. Final Activity Report No. 2015000094 (2018), p. 88. https://www.ctc-n.org/technical-assistance/projects/substantial-ghg-emissions-reduction-cement-industry-using-waste-heat

331. D. Sandalow et al., Carbon dioxide utilization roadmap 2.0 (ICEF, 2017), p. 30. https://www.icef.go.jp/platform/article_detail.php?article__id=171

332. H.J. Ho et al., CO2 utilization via direct aqueous carbonation of synthesized concrete fines under atmospheric pressure. ACS Omega **5**(26), 15877–15890 (2020). https://doi.org/10.1021/acsomega.0c00985

333. A. Iizuka et al., Pilot-scale operation of a concrete sludge recycling plant and simultaneous production of calcium carbonate. Chem. Eng. Commun. **204**(1), 79–85 (2017). https://doi.org/10.1080/00986445.2016.1235564

334. Y. Katsuyama et al., Development of a process for producing high-purity calcium carbonate (CaCO3) from waste cement using pressurized CO2. Environ. Prog. **24**(2), 162–170 (2005). https://doi.org/10.1002/ep.10080

335. D. Shuto et al., CO2 fixation process with waste cement powder via regeneration of alkali and acid by electrodialysis: effect of operation conditions. Ind. Eng. Chem. Res. **54**(25), 6569–6577 (2015). https://doi.org/10.1021/acs.iecr.5b00717

336. S. Hong et al., Integration of two waste streams for carbon storage and utilization: enhanced metal extraction from steel slag using biogenic volatile organic acids. ACS Sustain. Chem. Eng. **8**(50), 18519–18527 (2020). https://doi.org/10.1021/acssuschemeng.0c06355

337. N.L. Ukwattage et al., Steel-making slag for mineral sequestration of carbon dioxide by accelerated carbonation. Meas.: J. Int. Meas. Confed. **97**, 15–22 (2017). https://doi.org/10.1016/j.measurement.2016.10.057

338. C. Wei et al., Kinetics model adaptability analysis of CO2 sequestration process utilizing steel making slag and cold-rolling wastewater. J. Hazard. Mater. **404**, 124094, 9 (2021). https://doi.org/10.1016/j.jhazmat.2020.124094

339. H.J. Ho et al., Chemical recycling and use of various types of concrete waste: a review. J. Clean. Prod. **284**, 124785, 14 (2021). https://doi.org/10.1016/j.jclepro.2020.124785

340. S.K. Kaliyavaradhan et al., CO2 sequestration of fresh concrete slurry waste: optimization of CO2 uptake and feasible use as a potential cement binder. J. CO2 Util. **42**, 101330, 11 (2020). https://doi.org/10.1016/j.jcou.2020.101330

341. S.K. Kaliyavaradhan et al., Valorization of waste powders from cement-concrete life cycle: a pathway to circular future. J. Clean. Prod. **268**, 122358, 25 (2020). https://doi.org/10.1016/j.jclepro.2020.122358

342. L. Ji et al., Insights into carbonation kinetics of fly ash from victorian lignite for CO2 sequestration. Energy Fuels **32**(4), 4569–4578 (2018). https://doi.org/10.1021/acs.energyfuels.7b03137

343. W. Liu et al., CO2 sequestration by direct gas–solid carbonation of fly ash with steam addition. J. Clean. Prod. **178**, 98–107 (2018). https://doi.org/10.1016/j.jclepro.2017.12.281

344. Y. Izumi et al., Calculation of greenhouse gas emissions for a carbon recycling system using mineral carbon capture and utilization technology in the cement industry. J. Clean. Prod. **312**, 127618, 11 (2021). https://doi.org/10.1016/j.jclepro.2021.127618

345. M.D. Obrist et al., Decarbonization pathways of the swiss cement industry towards net zero emissions. J. Clean. Prod. **288**, 125413, 13 (2021). https://doi.org/10.1016/j.jclepro.2020. 125413

346. S. Pamenter, R.J. Myers, Decarbonizing the cementitious materials cycle: a whole-systems review of measures to decarbonize the cement supply chain in the UK and european contexts. J. Ind. Ecol. **25**(2), 359–376 (2021). https://doi.org/10.1111/jiec.13105

347. O. Ashrafi et al., Impact of carbon capture technologies on GHG emissions from oil sands in-situ facilities: a system prospective. Appl. Therm. Eng. **188**, 116603, 13 (2021). https://doi.org/10.1016/j.applthermaleng.2021.116603

348. H. Li et al., CO2 storage potential in major oil and gas reservoirs in the northern South China sea. Int. J. Greenh. Gas Control. **108**, 103328, 13 (2021). https://doi.org/10.1016/j.ijggc.2021. 103328

349. B. Yang et al., Life cycle cost assessment of biomass co-firing power plants with CO2 capture and storage considering multiple incentives. Energy Econ. **96** (2021). https://doi.org/10.1016/j.eneco.2021.105173

350. Office of Air Quality Planning and Standards, *Available and Emerging Technologies for Reducing Greenhouse Gas Emissions from the Petroleum Refining Industry* (US-EPA, Research Triangle Park, North Carolina, USA, 2010), p. 42

351. D. Shaw, L. Thomas, *Towards your low-carbon refinery* (Unpublished Presentation) (2019), p. 42. https://www.digitalrefining.com/news/1005678/towards-your-low-carbon-refinery#.YQ1liERKhPY

352. D. Chakrabarti et al., Conversion of CO2 over a co-based Fischer-Tropsch catalyst. Ind. Eng. Chem. Res. **54**(4), 1189–1196 (2015). https://doi.org/10.1021/ie503496m

353. P. Styring, G.R.M. Dowson, Oxygenated transport fuels from carbon dioxide. Johnson Matthey Technol. Rev. **65**(2), 170–179 (2021). https://doi.org/10.1595/205651321x16063027 322661

354. P.J. Hall et al., CO2-derived fuels for energy storage, in *Carbon Dioxide Utiiisation: Closing the Carbon Cycle*, ed. by P. Styring et al. (Elsevier, Amsterdam, 2015), pp. 33–44

355. N. Hernandez, La descarbonización del sistema mundial de transporte vehicular. Gerencia y Energia 15 (2021). https://www.scribd.com/document/520242204/La-Descarbonizacion-Del-Sistema-Mundial-de-Transporte-Terrestre. Accessed Aug 2021

356. R. Zhang et al., Long-term implications of electric vehicle penetration in urban decarbonization scenarios: an integrated land use–transport–energy model. Sustain. Cities Soc. **68**, 102800, 11 (2021). https://doi.org/10.1016/j.scs.2021.102800

357. T. Ayvalı et al., The position of ammonia in decarbonising maritime industry: an overview and perspectives: Part I. Johnson Matthey Technol. Rev. **65**(2), 275–290 (2021)

358. T. Ayvalı et al., The position of ammonia in decarbonising maritime industry: an overview and perspectives: Part II. Johnson Matthey Technol. Rev. **65**(2), 291–300 (2021)

359. F.Y. Al-Aboosi et al., Renewable ammonia as an alternative fuel for the shipping industry. Curr. Opin. Chem. Eng. **31** (2021). https://doi.org/10.1016/j.coche.2021.100670

360. P. Balcombe et al., How can LNG-fuelled ships meet decarbonisation targets? An environmental and economic analysis. Energy **227**, 120462, 12 (2021). https://doi.org/10.1016/j.energy.2021.120462

361. A. Fan et al., Decarbonising inland ship power system: Alternative solution and assessment method. Energy **226** (2021). https://doi.org/10.1016/j.energy.2021.120266

362. G. Mallouppas, E.A. Yfantis, Decarbonization in shipping industry: a review of research, technology development, and innovation proposals. J. Mar. Sci. Eng. **9**(4) 2021. https://doi.org/10.3390/jmse9040415

363. P. Chandhoke, Decarbonizing aviation. Sustainable growth with renewable jet fuel. Neste— Sustainable Jet Fuel (2021), p. 10. https://www.neste.com/what-is-neste-my-renewable-jet-fuel

364. P. Chandhoke, Renewable jet fuel, why does it cost more? (2018). https://www.neste.us/about-neste/news-inspiration/blog/aviation/renewable-jet-fuel-why-does-it-cost-more. Accessed July

365. A. Sherry et al., How to access and exploit natural resources sustainably: petroleum biotechnology. Microb. Biotechnol. **10**(5), 1206–1211 (2017). https://doi.org/10.1111/1751-7915. 12793

366. A. Sánchez et al., Evaluating ammonia as green fuel for power generation: a thermo-chemical perspective. Appl. Energy **293**, 116956, 12 (2021). https://doi.org/10.1016/j.apenergy.2021. 116956

367. P.P. Pichler et al., International comparison of health care carbon footprints. Environ. Res. Lett. **14**(6) (2019). https://doi.org/10.1088/1748-9326/ab19e1

368. N. Watts et al., Health and climate change: policy responses to protect public health. Lancet **386**(10006), 1861–1914 (2015). https://doi.org/10.1016/S0140-6736(15)60854-6

369. E. Prisman et al., Modified oxygen mask to induce target levels of hyperoxia and hypercarbia during radiotherapy: a more effective alternative to carbogen. Int. J. Radiat. Biol. **83**(7), 457–462 (2007). https://doi.org/10.1080/09553000701370894

370. L.J. Meduna, Autobiography of L. J. Meduna. Part I and II. Convuls. Ther. **1**(1–2), 43–57, 121–135 (1985)

371. H. Gadani, A. Vyas, Anesthetic gases and global warming: potentials, prevention and future of anesthesia. Anesth Essays Res **5**(1), 5–10 (2011). https://doi.org/10.4103/0259-1162.84171

372. D. Bennet et al., Evaluation of supercritical CO2 sterilization efficacy for sanitizing personal protective equipment from the coronavirus sars-cov-2. Sci. Total Environ. **780** (2021). https://doi.org/10.1016/j.scitotenv.2021.146519

373. M. Persson, J. Van Der Linden, The potential use of carbon dioxide as a carrier gas for drug delivery into open wounds. Med. Hypotheses **72**(2), 121–124 (2009). https://doi.org/10.1016/j.mehy.2008.08.026

374. M.I. Canto et al., Carbon dioxide (CO2) cryotherapy is a safe and effective treatment of barrett's esophagus (be) with hgd/intramucosal carcinoma. Gastrointest. Endosc. **69**(5), AB341 (2009). https://doi.org/10.1016/j.gie.2009.03.994

375. J.E.R.E. Wong Chung et al. CO2-lasertonsillotomy under local anesthesia in adults. JoVE (153), e59702, 7 (2019). https://doi.org/10.3791/59702

376. S. Søvik, K. Lossius, Development of ventilatory response to transient hypercapnia and hypercapnic hypoxia in term infants. Pediatr. Res. **55**(2), 302–309 (2004). https://doi.org/10.1203/01.PDR.0000106316.40213.DB

377. M.V. Natu, H.A. Avery, Supercritical CO2 encapsulation of cosmetic ingredients: novel methods for tailoring ingredients for the cosmetics industry. Househ. Pers. Care Today **93**(3), 42–45 (2014)

378. F. Temelli, Perspectives on the use of supercritical particle formation technologies for food ingredients. J. Supercrit. Fluids **134**, 244–251 (2018). https://doi.org/10.1016/j.supflu.2017. 11.010

379. V. Nedovic et al., An overview of encapsulation technologies for food applications. Procedia Food Sci. **1**, 1806–1815 (2011). https://doi.org/10.1016/j.profoo.2011.09.265

380. B.F. Gibbs et al., Encapsulation in the food industry: a review. Int. J. Food Sci. Nutr. **50**(3), 213–224 (1999). https://doi.org/10.1080/096374899101256

381. S. Klettenhammer et al., Novel technologies based on supercritical fluids for the encapsulation of food grade bioactive compounds. Foods **9**(10) (2020). https://doi.org/10.3390/foods9 101395

382. S.C. Lourenço et al., Antioxidants of natural plant origins: from sources to food industry applications. Molecules **24**(22) (2019). https://doi.org/10.3390/molecules24224132

383. N.E. Polyakov, T.V. Leshina, *Certain aspects of the reactivity of carotenoids. Redox processes and complexation.* Russ. Chem. Rev. **75**(12), 1049–1064 (2006). https://doi.org/10.1070/RC2 006v075n12ABEH003640

384. T. Sen et al., Microbial pigments in the food industry—challenges and the way forward. Front. Nutr. **6**(2019). https://doi.org/10.3389/fnut.2019.00007

385. P. Kaushik et al., Microencapsulation of omega-3 fatty acids: a review of microencapsulation and characterization methods. J. Funct. Foods **19**, 868–881 (2015). https://doi.org/10.1016/j. jff.2014.06.029

386. L. Zhao et al., Encapsulation of lutein in liposomes using supercritical carbon dioxide. Food Res. Int. **100**, 168–179 (2017). https://doi.org/10.1016/j.foodres.2017.06.055

387. R. Thiering et al., Current issues relating to anti-solvent micronisation techniques and their extension to industrial scales. J. Supercrit. Fluids **21**(2), 159–177 (2001). https://doi.org/10.1016/S0896-8446(01)00090-0

388. S. Varona et al., Formulation of lavandin essential oil with biopolymers by pgss for application as biocide in ecological agriculture. J. Supercrit. Fluids **54**(3), 369–377 (2010). https://doi.org/10.1016/j.supflu.2010.05.019

389. A. Visentin et al., Precipitation and encapsulation of rosemary antioxidants by supercritical antisolvent process. J. Food Eng. **109**(1), 9–15 (2012). https://doi.org/10.1016/j.jfoodeng.2011.10.015

390. O. Yesil-Celiktas, E.O. Cetin-Uyanikgil, In vitro release kinetics of polycaprolactone encapsulated plant extract fabricated by supercritical antisolvent process and solvent evaporation method. J. Supercrit. Fluids **62**, 219–225 (2012). https://doi.org/10.1016/j.supflu.2011.11.005

391. D. Hu et al., Preparation, characterization, and in vitro release investigation of lutein/zein nanoparticles via solution enhanced dispersion by supercritical fluids. J. Food Eng. **109**(3), 545–552 (2012). https://doi.org/10.1016/j.jfoodeng.2011.10.025

392. B.K. Sovacool et al., Decarbonizing the food and beverages industry: a critical and systematic review of developments, sociotechnical systems and policy options. Renew. Sustain. Energy Rev. **143** (2021). https://doi.org/10.1016/j.rser.2021.110856

393. M. Schipek, Treatment of acid mine lakes—lab and field studies. PhD Thesis from Freiberg Online Geology, 2011, pp. 381.

394. M. Werner et al., Carbonation of activated serpentine for direct flue gas mineralization, in *Proceedings of the 11th International Conference on Greenhouse Gas Control Technologies, GHGT 2012*, vol. 37 (Elsevier Ltd, Kyoto, 2013), pp. 5929–5937. https://doi.org/10.1016/j.egypro.2013.06.519

395. M. Nagy et al., Characterization of CO2 precipitated kraft lignin to promote its utilization. Green Chem. **12**(1), 31–34 (2010)

396. R. Xing et al., Deep decarbonization pathways in the building sector: China's NDC and the Paris agreement. Environ. Res. Lett. **16**(4) (2021). https://doi.org/10.1088/1748-9326/abe008

397. P. Aparicio et al., Behaviour of concrete and cement in carbon dioxide sequestration by mineral carbonation processes. Boletin de la Sociedad Espanola de Ceramica y Vidrio 255, 9 (2020). https://doi.org/10.1016/j.bsecv.2020.11.003

398. M. Kueppers et al., *Decarbonization pathways of worldwide energy systems—definition and modeling of archetypes.* Appl. Energy **285** (2021). https://doi.org/10.1016/j.apenergy.2021.116438

399. National Academies of Sciences-Engineering-Medicine, *Accelerating decarbonization of the U.S. Energy System* (The National Academies Press, Washington, DC, 2021), p. 268. https://doi.org/10.17226/25932

400. J.A. Rodríguez-Sarasty et al., Deep decarbonization in northeastern North America: the value of electricity market integration and hydropower. Energy Policy. **152** (2021). https://doi.org/10.1016/j.enpol.2021.112210

401. J.W. Dijkstra et al., Techno-economic evaluation of membrane technology for pre-combustion decarbonisation: water-gas shift versus reforming. Energy Procedia **4**, 723–730 (2011). https://doi.org/10.1016/j.egypro.2011.01.111

402. K. Kugeler et al., Transport of nuclear heat by means of chemical energy (nuclear long-distance energy). Nucl. Eng. Des. **34**(1), 65–72 (1975). https://doi.org/10.1016/0029-5493(75)90156-9

403. C. Vogt et al., The renaissance of the sabatier reaction and its applications on earth and in space. Nat. Catal. **2**(3), 188–197 (2019). https://doi.org/10.1038/s41929-019-0244-4

404. S. Schiebahn et al., Power to gas: technological overview, systems analysis and economic assessment for a case study in germany. Int. J. Hydrogen Energy **40**(12), 4285–4294 (2015). https://doi.org/10.1016/j.ijhydene.2015.01.123

405. J.T. Richardson, S.A. Paripatyadar, Carbon dioxide reforming of methane with supported rhodium. Appl. Catal. **61**(1), 293–309 (1990). https://doi.org/10.1016/S0166-9834(00)82152-1

406. J.H. Mccrary et al., An experimental study of the CO2CH4 reforming-methanation cycle as a mechanism for converting and transporting solar energy. Sol. Energy **29**(2), 141–151 (1982). https://doi.org/10.1016/0038-092X(82)90176-1

407. T.A. Chubb et al., Loss factors in the design of thermochemical power plants, CO2-CH4 vs. SO3 chemical transplant fluids, in *Sun 2, Proceedings of the International Solar Energy Society. Silver Jubilee Congress* (1979), pp. 126–129

408. T.A. Chubb et al., Application of chemical engineering to large scale solar energy. Sol. Energy **20**(3), 219–224 (1978). https://doi.org/10.1016/0038-092X(78)90100-7

409. R.B. Diver, Transporting solar energy with chemistry. ChemTech **17**(10), 606–611 (1987)

410. J.D. Fish, D.C. Hawn, Closed loop thermochemical energy transport based on CO2 reforming of methane: Balancing the reaction systems, in *Proceedings of the 21st Intersociety Energy Conversion Engineering Conference: Advancing toward Technology Breakout in Energy Conversion* (ACS, Washington, DC, USA. San Diego, CA, USA, 1986), pp. 935–940

411. V.I. Anikeev et al., Experimental study of a catalytic solar energy device based on a closed thermochemical cycle. Dokl. Chem. Technol. **292–294**, 30–35 (1987)

412. V.I. Anikeev et al., Thermocatalytic solar-to-chemical energy transducer with a high energy-storage coefficient. Dokl. Chem. Technol. **289–291**, 85–89 (1986)

413. T.A. Chubb, Solar thermochemical energy. European Space Agency, (Special Publication) ESA SP **1**, 293–295 (1979)

414. T.A. Chubb et al., Design of a small thermochemical receiver for solar thermal power. Sol. Energy **23**(3), 217–221 (1979). https://doi.org/10.1016/0038-092X(79)90161-0

415. T. Chubb, A chemical approach to solar energy. Chem. Tech. **6**, 654–657 (1976)

416. G. De Maria et al., Thermochemical storage of solar energy with high-temperature chemical reactions. Sol. Energy **35**(5), 409–416 (1985). https://doi.org/10.1016/0038-092X(85)90129-X

417. R. Levitan et al., Closed-loop operation of a solar chemical heat pipe at the weizmann institute solar furnace. Sol. Energy Mater. **24**(1–4), 464–477 (1991). https://doi.org/10.1016/0165-1633(91)90083-W

418. R. Rubin et al., Methanation of synthesis gas in a solar chemical heat pipe. Energy **17**(12), 1109–1119 (1992). https://doi.org/10.1016/0360-5442(92)90001-G

419. F. Safari, I. Dincer, Assessment and optimization of an integrated wind power system for hydrogen and methane production. Energy Convers. Manage. **177**, 693–703 (2018). https://doi.org/10.1016/j.enconman.2018.09.071

420. A. Segal, M. Levy, Solar chemical heat pipe in closed loop operation: mathematical model and experiments. Sol. Energy **51**(5), 367–376 (1993). https://doi.org/10.1016/0038-092X(93)90149-I

421. G.A. Olah, Beyond oil and gas: the methanol economy. Angew. Chem.-Int. Ed **44**(18), 2636–2639 (2005). https://doi.org/10.1002/anie.200462121

422. A. Francis et al., A review on recent developments in solar photoreactors for carbon dioxide conversion to fuels. J. CO2 Util. **47**, #101515, 19 (2021). https://doi.org/10.1016/j.jcou.2021.101515

423. R. Snoeckx, A. Bogaerts, Plasma technology-a novel solution for CO2 conversion? Chem. Soc. Rev. **46**(19), 5805–5863 (2017). https://doi.org/10.1039/c6cs00066e

424. M. Mikhail et al., Plasma-catalytic hybrid process for CO2 methanation: optimization of operation parameters. React. Kinet. Mech. Catal. 15 (2018). https://doi.org/10.1007/s11144-018-1508-8

425. D. Yıldız, How much will the renewables help for decarbonised future? World Water Dipl. Sci. News 10002, 16 (2018)

426. J. Lattner, S. Stevenson, Renewable power for carbon dioxide mitigation. Chem. Eng. Prog. 54–58 (2021)

427. B. Guler et al., Cost-effective decarbonization through investment in transmission interconnectors as part of regional energy hubs (REH). Electr. J. **34**(3), 106924, 10 (2021). https://doi.org/10.1016/j.tej.2021.106924

428. H.M. Kvamsdal et al., Exergy analysis of gas-turbine combined cycle with CO2 capture using pre-combustion decarbonization of natural gas, in *Proceedings of the American Society of Mechanical Engineers*, vol. **2B** (International Gas Turbine Institute, Turbo Expo, 2002), pp. 675–682. https://doi.org/10.1115/GT2002-30411

429. The Lancet Planetary Health, Net zero clarity. Lancet Planet. Health **5**(4), e176 (2021). https://doi.org/10.1016/S2542-5196(21)00054-1

430. United Nations, *Glasgow climate pact*. COP26: Decision-/CMA.3 (2021), p. 11. https://unfccc.int/sites/default/files/resource/cma3_auv_2_cover%2520decision.pdf. Accessed Dec 2021

431. Editorial, Net-zero carbon pledges must be meaningful to avert climate disaster. Nature **592**(7852), 8 (2021). https://doi.org/10.1038/d41586-021-00864-9

432. H.L. Van Soest et al., Net-zero emission targets for major emitting countries consistent with the Paris agreement. Nat. Commun. **12**(1) (2021). https://doi.org/10.1038/s41467-021-22294-x

433. J. Rogelj et al., Scenarios towards limiting global mean temperature increase below 1.5 °C. Nat. Clim. Chang. **8**(4), 325–332 (2018). https://doi.org/10.1038/s41558-018-0091-3

434. D. Huppmann et al., IAMC 1.5 °C scenario explorer and data hosted by iiasa. Version 1.1 Integrated Assessment Modeling Consortium & International Institute for Applied Systems Analysis 2018. https://doi.org/10.22022/SR15/08-2018.15429. https://data.ene.iiasa.ac.at/DOI/SR15/08-2018.15429/. Accessed Mar 2022

435. G.P. Thiel, A.K. Stark, To decarbonize industry, we must decarbonize heat. Joule **5**(3), 531–550 (2021). https://doi.org/10.1016/j.joule.2020.12.007

436. M. Allen et al., Global warming of 1.5 °C. Special Report. Intergovernmental Panel on Climate Change (IPCC). (Cambridge University Press, UK, 2018), pp. 540. https://www.ipcc.ch/sr15/

437. J. Ranganathan et al., The greenhouse gas protocol: A corporate accounting and reporting standard. (World Resources Institute, World Business Council for Sustainable Development, USA, 2004), p. 116. https://ghgprotocol.org/corporate-standards

438. S. Hill et al., Net zero carbon buildings: three steps to take now. ARUP. (2020), p. 14

439. R. Cohen et al., Net zero carbon: energy performance targets for offices. Build. Serv. Eng. Res. Technol. **42**(3), 349–369 (2021). https://doi.org/10.1177/0143624421991470

440. H. Bahar et al., Renewables 2019 (IEA, Paris, France, 2019), p. 204. https://www.iea.org/reports/renewables-2019

441. P.C. Slorach, L. Stamford, Net zero in the heating sector: Technological options and environmental sustainability from now to 2050. Energy Convers. Manag. **230**(2021). https://doi.org/10.1016/j.enconman.2021.113838

442. W. Wei, H.M. Skye, Residential net-zero energy buildings: review and perspective. Renew. Sustain. Energy Rev. **142**(2021). https://doi.org/10.1016/j.rser.2021.110859

443. V. Camobreco et al., Inventory of U.S. Greenhouse gas emissions and sinks: 1990–2019. EPA 430-R-21–005 Report (US Environmental Protection Agency, Washington, DC, USA, 2021), p. 790. https://www.epa.gov/ghgemissions/inventory-us-greenhouse-gas-emissions-and-sinks

444. J. Pedraza et al., On the road to net zero-emission cement: Integrated assessment of mineral carbonation of cement kiln dust. Chem. Eng. J. **408**(2021). https://doi.org/10.1016/j.cej.2020.127346

445. J. Morfeldt et al., Carbon footprint impacts of banning cars with internal combustion engines. Transp. Res. Part D: Transp. Environ. **95**, 102807, 19 (2021). https://doi.org/10.1016/j.trd.2021.102807

446. J. Su et al., Challenges in determining the renewable content of the final fuels after co-processing biogenic feedstocks in the fluid catalytic cracker (FCC) of a commercial oil refinery. Fuel **294**, 120526, 9 (2021). https://doi.org/10.1016/j.fuel.2021.120526

447. V. Becattini et al., Role of carbon capture, storage, and utilization to enable a net-zero-CO2-emissions aviation sector. Ind. Eng. Chem. Res. (2021). https://doi.org/10.1021/acs.iecr.0c05392

448. D. Hall et al., Beyond road vehicles: Survey of zero-emission technology options across the transport sector. International Council on Clean Transportation Working Paper 2018 www.theicct.org, Working Paper 2018–11, p. 22

449. International Council on Clean Transportation, Aviation (2021). https://theicct.org/sector/aviation/. Accessed Mar 2023

450. O. Rueda et al., Negative-emissions technology portfolios to meet the 1.5 °C target. Glob. Environ. Chang. **67**, #102238, 13(2021). https://doi.org/10.1016/j.gloenvcha.2021.102238

451. E. Kato, A. Kurosawa, Role of negative emissions technologies (NETs) and innovative technologies in transition of japan's energy systems toward net-zero CO_2 emissions. Sustain. Sci. **16**(2), 463–475 (2021). https://doi.org/10.1007/s11625-021-00908-z

452. S. Roe et al., Contribution of the land sector to a 1.5 °C world. Nat. Clim. Chang. **9**(11), 817–828 (2019). https://doi.org/10.1038/s41558-019-0591-9

453. M.N. Mohd Idris et al., Spatio-temporal assessment of the impact of intensive palm oil-based bioenergy deployment on cross-sectoral energy decarbonization. Appl. Energy **285**(2021). https://doi.org/10.1016/j.apenergy.2021.116460

454. J. Monjardino et al., Carbon neutrality pathways effects on air pollutant emissions: the portuguese case. Atmosphere **12**(3), 18 (2021). https://doi.org/10.3390/atmos12030324

455. Y. Wang et al., Economic and land use impacts of net zero-emission target in new zealand. Int. J. Urban Sci. (2021). https://doi.org/10.1080/12265934.2020.1869582

456. S.K. Mahapatra et al., An assessment of factors contributing to firms' carbon footprint reduction efforts. Int. J. Prod. Econ. **235**, 108073, 11 (2021). https://doi.org/10.1016/j.ijpe.2021.108073

457. J. Kim et al., Computing a strategic decarbonization pathway: a chance-constrained equilibrium problem. IEEE Trans. Power Syst. **36**(3), 1910–1921 (2021). https://doi.org/10.1109/TPWRS.2020.3038840

458. C. Stark et al., *The Path to Net Zero* (Climate Assembly, London, UK, 2020), p. 556

459. R. Zoboli et al., Energy and the circular economy: Filling the gap through new business models within the EGD (Fondazione Eni Enrico Mattei, Sustainability, Environmental Economics, and Dynamics Studies. Milan, Italy, 2020), p. 35. https://www.feem.it/en/publications/reports/energy-and-the-circular-economy-filling-the-gap-through-new-business-models-within-the-egd/

460. C. Greig, Getting to net-zero emissions. Engineering **6**(12), 1341–1342 (2020). https://doi.org/10.1016/j.eng.2020.09.005

461. IPCC, Climate change and land. Special Report. Intergovernmental Panel on Climate Change (IPCC). (Cambridge University Press, UK, 2020), p. 540

462. A. Popp et al., Land-use futures in the shared socio-economic pathways. Glob. Environ. Chang. **42**, 331–345 (2017). https://doi.org/10.1016/j.gloenvcha.2016.10.002

463. A. Lamb et al., The potential for land sparing to offset greenhouse gas emissions from agriculture. Nat. Clim. Chang. **6**(5), 488–492 (2016). https://doi.org/10.1038/nclimate2910

464. P. Smith et al., Biophysical and economic limits to negative CO_2 emissions. Nat. Clim. Chang. **6**(1), 42–50 (2016). https://doi.org/10.1038/nclimate2870

Chapter 3
Sustainable Circularity

3.1 Definitions

The decarbonization of the C-emitting economic sectors is a mandatory need for GHGs reduction and control gain over climate change. As mentioned in previous chapters, not only new technologies are needed, but creation of public policies and incentives are also required. Additional requirements are implicitly subscribed, and this is to say that any solution must meet sustainability principles. Sustainable decarbonizing technologies are the best and most suitable options for integration within cyclic systems, to achieve sustainable circular economic solutions. This chapter discusses different aspects of a sustainable circular economy, starting with a set of definitions that set the context of the discussion, followed by reviewing the cumulative knowledge on strategies, approaches, and business models developed to close the cycle involving C-bearing compounds and materials.

A linear economy is characterized by coupled or integrated supply chains that take resources and employ energy to manufacture products/devices used by society that then disposes spent or wasted materials or products. In contrast, the circular economy (CE) emerges as an integral evolution for a better resource management and environment protection. Thus, the General Union Environment Action Programme of the European Union has defined the CE as one *"where nothing is wasted and where natural resources are managed sustainably, and biodiversity is protected, valued and restored in ways that enhance our society's resilience; and our low-carbon growth has long been decoupled from resource use, setting the pace for a safe and sustainable global society"* [1]. Thus, CE is described as an approach to promote the responsible and cyclical use of resources. Accordingly, the CE value chain was distinguished by a closed loop of material flow and was suggested to be driven by RE [2], but we are inclined to extend this proposition to include carbon–neutral energy resources. The environmental sustainability inspired within CE has to parallelize economic sustainability, by promoting economic growth. CE associated technologies are expected to create new businesses and job opportunities, save costs,

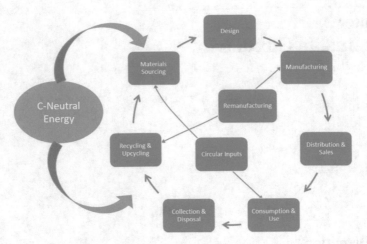

Fig. 3.1 The circular economy concept. (Reproduced from Ref. [4], with permission from Elsevier)

reduce price volatility, and improve supply security [3]. Therefore, carbon neutrality and even negative carbon pathways can achieve environmental, energy, and economic sustainability if conceived within a circular concept, as visualized in Fig. 3.1 [4]. Recycling has been the focus in most of the implemented CE strategies, but an assessment indicated that a single focus was not enough to reach sustainability [5], as it was also confirmed when the focus was waste processing [6]. A right metrics is still needed to evaluate circular strategies, for which new comprehensive methodologies have to be developed [5], involving all steps and aspects of Fig. 3.1, in a holistic manner. The success of any decarbonization strategy including CO_2 utilization or reuse must be assessed using such comprehensive methodology that appreciates the intrinsic value of CO_2 chemical conversion, within a CE concept.

An overview of the CE concept based on a revision of published definitions has been provided by Kalmykova et al. with the purposes of detecting commonalities in the concept and of deriving the underpinning principles [2]. These principles include (i) resource optimization, (ii) eco-efficiency, (iii) eco-effectiveness, (iv) waste prevention. Resource optimization is determined by the recognition that there is no unlimited source of materials. While eco-efficiency is the minimization of the volume, velocity, and toxicity of materials flowing through the systems, eco-effectiveness concerns a supportive relationship between the transformation of products and their associated material flows with ecological systems and future economic growth. Finally, waste prevention leads to the reduction, reuse, recycle, recover, and repurposing of materials and products.

Probably, the most complete graphical representation of the resource management within a CE has been illustrated by the Ellen MacArthur Foundation's "butterfly diagram" [7], in which each of the material fluxes, i.e., the "biological" (renewable) and the "fossil" flows in distinct cycles. However, the realization that in many instances resource flows comprise combinations of renewable and fossil materials,

Fig. 3.2 Integrated resource flow diagram. (Reproduced from Ref. [8], under unrestricted Open Access License Agreement)

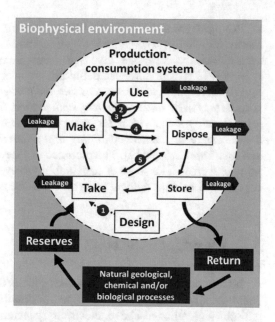

either naturally occurring materials or by technical design moves toward an integrated flow version, as the one proposed by Velenturf et al. [8] and showed in Fig. 3.2. This diagram (Fig. 3.2) reshapes the CE concept (Fig. 3.1) by broadening its scope and covering extractive sectors and the return of materials from the anthropogenic use to natural reserves. This more realistic representation including biophysical limitations and endeavors could be taken as the basis for a transition towards a sustainable CE [8].

Although emissions and wastes are avoided in the represented circularity, indicating it might be sustainable in some instances, only value chain cycles could support sustainability. Additionally, another implicit idea is the end-to-end principle supported by the self-regenerative nature of the cycles.

There are more than enough environmental arguments calling for the implementation of a sustainable CE. Sustainability was defined in 1987, by the UN Brundtland Commission as "*meeting the needs of the present without compromising the ability of future generations to meet their own needs*" [9]. This definition of sustainability implicitly involves three components: social, environment and economy. Therefore, both production and end-use of resources, matter, products, and energy have to be pursued within an environmental conscience and economically, to preserve the well-being of the future generations.

In 2015, the UN defined 17 Sustainable Development Goals (SDGs) for the 2030 Agenda [10], which were considered an urgent call for action by all countries in a global partnership. Within these SDGs, it was reasoned the key role of bioeconomy, in paving the way for achieving the goals. From the global perspective, the International Advisory Council defined thus bioeconomy, in broad, diverse, and general terms as "*the production, utilization and conservation of biological resources, including related knowledge, science, technology, and innovation, to provide information, products, processes and services across all economic sectors aiming toward a sustainable economy*" [11]. The European definition is somehow more specific: "*the production of renewable biological resources and the conversion of these resources and waste streams into value added products, such as food, feed, bio-based products and bioenergy. Its sectors and industries have strong innovation potential due to their use of a wide range of sciences, enabling and industrial technologies, along with local and tacit knowledge*" [12]. The presence of renewable resources and environmental principles provides a sense of sustainability. Nevertheless, it does not seem to be the case, as will become evident from the next discussion.

3.2   Closing the Carbon Cycle

As pointed out above, a single approach such as the transport electrification strategy cannot be taken as a universal solution, for decarbonization and climate change mitigation. As mentioned above, electricity generation does represent a large proportion of GHG emissions [13]. In fact, the electricity supply sector is responsible for nearly 40% of the global CO_2 emissions, accounting for the release of over 7700 Mton$_{CO2}$/y [14]. Therefore, unless the electric sector firstly passes through a drastic decarbonization, any electrification strategy will fail or be too slow to achieve the sustainability goals. The evolution of the smart grid concept into the smart energy concept, expanding from transport and distribution to also include generation and storage, will facilitate the decarbonization of electrification. An increase in security while decreasing environmental impact and cost have been pointed out as the main benefits of the smart energy concept [15].

The need for an increase in renewable or low-carbon energy that provokes substantial reduction of the CO_2 emissions per kWh is evident and urgent. Starting with the transport sector, the use of biofuels and particularly, second-generation biofuels is considered an option for the significant reduction of life cycle emissions since these are produced from the whole plant, agricultural and forestry residues, and renewable wastes. The Environment Program (UNEP) of the UN Trade and Agriculture Directorate compared the overall GHG mitigation potential of biofuels to that of fossil fuels, from the data and results of 60 published lifecycle analysis studies. The results demonstrate the key role of biofuels in any decarbonizing strategy [16]. The increase in renewable or low-carbon energy, nonetheless, has a limit. A recent study rose concerns on the potential for RE to become the prominent source of primary energy. The fact is that under the business-as-usual scenario, global energy demand will increase to 1000EJ, by 2050. An analysis considering economic factors and climate change effects on REs (tidal, geothermal, solar, wind, and hydropower) indicated their unlikely potential for fulfilling the energy demand [17]. An enormous increase in energy efficiency is required to compensate the increase in energy demand, but also cultural changes are needed to decrease the social thirst for energy, if environmental sustainability is appreciated at the horizon.

Carbon management within a circular or closed-loop economy leads to the concept of carbon conservation. Carbon is essential to life on Earth and, therefore, its conservation is vital in preserving the nature from a catastrophic collapse. Thus, carbon conservation is practiced by a careful management of the carbon-bearing species, compounds, and materials, preferentially through a CE that includes CC, CS, and CU technologies. However, it has been recognized that only through utilization significant contribution to the CE will be made [18]. In fact, current carbon capture and storage (CCS) technologies have not resulted as cost-effective as desired since their cost is high, and their long-term sequestration effectiveness has not been proven. Therefore, cost-effective technologies for the valorization of CO_2 are needed to turn around the economy of CCU, in sustainable terms. Within the purpose of this work, environmental sustainability can only be achieved when the carbon cycle is closed efficiently. The context of efficiency has also to be expanded throughout environment and energy, i.e., eco-efficiency.

An emphasis is made once more on that today's linear "take, make, dispose" economy needs to be replaced by an eco-efficient closed-loop approach to production processes. CE is an urgent need and cannot be prolonged into a future prospect. New eco-efficient chemical processes enabling reuse of carbon sources are needed. These processes will use waste as feedstocks, becoming a way of feedstock recycling, which could turn difficult-to-recycle wastes into valuable chemical building blocks [18].

In the case of CO_2, the concept of a biorefinery as a way of making CCU for the production of fuels and bulk chemicals financially attractive [19] could turn CCU environmentally beneficial only through eco-efficient processes. This CO_2 biorefinery concept was centered on two core pathways described as Schemes 1 and 2 that lead to the production of fuels and chemicals, respectively [19]. However, to achieve sustainability, any CO_2 biorefinery concept needs a suite of integrated (eco-efficient and eco-designed) process technologies that also fulfill the net-zero goals.

Biological valorization of CO_2 has been suggested as the basis for an environmentally and economically sustainable technology. Microalgae was considered an option to include effective capture and photosynthetic conversion of CO_2 into lipids (and other algal products) [20]. However, the current status of microalgae-derived process technologies, for the bioconversion of CO_2 remains far from economically, environmentally, and energy sustainable.

So far, the C-loop remains open and needs to be closed efficiently, as mentioned above. CO_2 is emitted from mobile or static sources and may or may not be captured. The conceptual carbon cycle described and discussed in Ref. [21], which could start with the CO_2 emissions from the application of a process at industry or at a power plant. These emissions should be directly captured at emission location. The captured-CO_2 could be stored via sequestration or converted into a product with longer life cycle. Both capture and conversion may or may not require to be separate steps, but they certainly need additional elements to fulfill operations and these additional elements are energy and raw materials or co-reagents. Figure 3.3 illustrates the carbon management operations involved in such C-cycle scheme, showing in green in the cycle steps. This cycle can be seen as a whole C-supply chain but also as the society engagement in the processes. Figure 3.3 also shows that CO_2, whether captured in a product, converted, or utilized would get eventually re-emitted from the use or consumption of the product by the end-user. The time CO_2 spends in the product depends on the type of produced material and on its application. Two extreme scenarios were discussed in [21]. The case of polymers or inorganic carbonates (cement, aggregates) make life cycle longer. Further, polymers may be recycled, reprocessed, or remanufactured, increasing the lifetime even more. Alternatively, conversion to fuels would lead to combustion and fast re-emission into the atmosphere, from where capturing is technologically limited. As can be seen, the most important candidates for capturing units are stationary sources from industrial processes. In this drawing, the potential CO_2 emissions are included in blue squares, and the red marked squares identify energy consuming steps that might become attention points. In fact, CO_2 processing typically involves high energy consuming processes, due to their low reactivity and high stability [4]. The needs for intensification between the capturing and utilization steps become evident, from observing this figure (Fig. 3.3) and would satisfy the previously mentioned requirements for increased energy-efficient technologies. Finally, when the co-reagent is water, both reactants could be captured from the atmosphere [22] and the reactions might be inspired on photosynthetic routes to mimic nature [23]. However, the energy intensity of the water-CO_2 reactions calls for low-carbon or RE sources, and also more energy-efficient chemical processes. If these challenges are overcome, new process technologies could be developed for sustaining a C-based circular economy [24].

The environmental (emissions) and energy intensity of each component of the C-cycle needs to be considered individually, integrally, and holistically. Additionally, the economy intensity will ultimately define feasibility. Since CCU, per se, cannot be

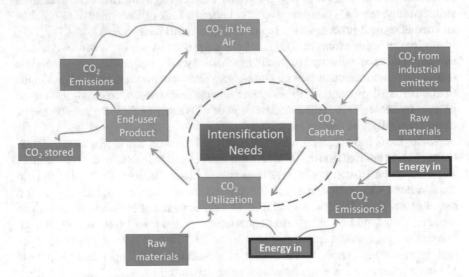

Fig. 3.3 Schematic representation of the C-cycle

taken as environmentally beneficial, LCA is of remarkable importance within the C-cycle analysis. LCA assesses environmental impacts by evaluating the entire life cycle of a product from cradle to grave. Each of the steps considered in the scheme shown in Fig. 3.3 involves energy demands and emissions. CO_2 is an energy demanding feedstock that requires large amount of energy for its activation. In some instances, H_2 is used as a co-feed or co-reagent, in which case CO_2 emissions are collateral to H_2 production. The capture, compression, and transport steps also involve energy that might have associated CO_2 emissions. It has been emphasized that most of the time the avoided CO_2 emissions by CO_2 capture are smaller than the amount of CO_2 captured. The need for standardization of LCA tools and methods is also noticed when comparing reported analyses [25].

The relevance of LCA in the significance of the closure of the C-cycle underpinned the impact of the circular economy (CE) on carbon emissions from the coal-fueled power generation industry in China. LCA was used to build an emissions reduction matrix for the carbon footprint of that industry. The quantitative analysis of the CE impact on industry indicated that resources recycling could contribute to the replacement of carbon-intense energy sources. The reduction of the carbon footprint due to direct emissions was estimated to be in the order of 20%. The model allowed the quantitative evaluation of indirect carbon emissions caused by the CE and determined the correlation between CE and carbon emissions reduction [26]. CCS and CCU technologies are needed for providing the desired environmental benefits. However, any emerging technology must be thoroughly evaluated, and LCA has been widely and successfully applied for such purpose [27]. The decarbonization principles that should be included within any LCA are illustrated in Fig. 3.4. These principles include the completeness of both lifecycle (organizations and products) as well as all GHGs besides CO_2, in which case CO_2-equivalent could be used as content proxy; comprehensive and consistent assessment of the decarbonization strategies should include the entire set of connected environmental involvements and should avoid any trade-offs; rather than incorporation of economic instruments or virtual baseline scenarios, reducing GHG impacts is preferred, particularly those which could be supported by evidence; and the calculation of carbon footprint should be kept and reported distinctly from the offsets accounting [28].

An attempt to evaluate the overall CO_2 balance of a new low-C process, for the production of a new construction material used LCA for the assessment of the GHGs balance and of the resource efficiency, in terms of water and energy use. The process design intention was at least C-neutrality though the ultimate target was negative carbon. The main component of the construction material was nesquehonite (magnesium carbonate trihydrate), which was synthesized from an aqueous precipitation of desalination brines, using an alkaline agent. The amount of alkali consumed, and its precipitation efficiency strongly affected the CO_2 emissions. A negative footprint could only be achieved if the process emissions reach 22–27 $kg_{CO2}/kmol_{alkalinity}$. Since no commercial alkali source reaches such levels of associated emissions, there will not be any alkaline wastes that could be used [29].

One of the preferred products from the water-CO_2 reactions is synthesis gas or syngas, a mixture of carbon monoxide and hydrogen, from which a variety of

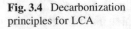
Fig. 3.4 Decarbonization
principles for LCA

hydrocarbons and oxygenates can be produced. The economical evaluation of syngas
production from three different pathways has been reported. The pathways considered the reduction of CO_2, from direct air capture (DAC) to syngas plants, via (i)
thermocatalytic reverse water gas shift (RWGS, see reaction 16, in Chap. 2); (ii)
gaseous CO_2-electrolysis (see Sect. 2.3.1.3. Co-electrolysis of CO_2 and water, in
Chap. 2), and (iii) direct (bi)carbonate electrolysis of the carbon capture solution.
Although at higher production costs (1.90 \$/kg), the DAC-(bi)carbonate electrolysis
offered lowest net CO_2 emissions. The gaseous electrolysis pathway produced syngas
at 1.30 \$/kg while the best cost (1.1 \$/kg) was obtained for the conventional rWGS
pathways. Any of the electrochemical pathways requires significant improvements
in performance, mainly FE to CO and cell voltage [30].

The U.S. National Academy of Science, Engineering, and Medicine has reviewed
the status of technologies for the utilization of gaseous carbon waste streams and
derived the research needs for the development and deployment of such processes.
Both achieving environmental goals and creating revenue from making carbon-based
products were targeted. Three pathways were considered, namely, mineralization,
chemical conversion, and biological processing. Although significant resources have
been invested nationally and internationally in this area, very little coordination, integration, and collaborative efforts were identified. The need for new processes and
products unveiled R&D gaps in areas such as new materials for additive manufacturing of process components, mapping and characterization of resources, pathways to new products, new catalysts and their integration to reactor design, system
engineering, bioreactors, and bioprospecting, for instance [31]. Another report was
devoted to technologies with negative emissions, i.e., those that remove and sequester
carbon dioxide. The needs for covering aspects from fundamental science to full-
scale deployment, as well as assessing the benefits, risks, and "sustainable scale
potential" were reported [32].

Regarding energy intensity, the accelerated development and deployment of low-carbon generation infrastructure is imperative. However, environmental sustainability of the generation infrastructure, supported by favorable LCA should be involved. A holistic view of the materials and components, including their extraction or manufacturing needs to be incorporated under the principles of the circular economy, with sustainable criteria on durability, reuse, and remanufacturing. Currently, the low-carbon energy sector is extracting, processing, and deploying millions of tons of composites, precious and rare earth materials throughout their infrastructure. There are no indications on these materials being recovered, managed, and returned to productive use sustainably, at a reasonable scale for enabling accelerated deployment [33].

The effective closure of the C-cycle through CCU through valorization requires the articulation of several strategies, some of which have already been mentioned, e.g., implementation of national sustainable policy, extracting valuable materials and elements from wastes, improved pollutants removal from flue gases, generation of high-value added products for diversified applications, integrated approach to multi-waste treatment as green solutions, and business models based on industrial eco-designed parks. However, only if the goal of zero waste is incorporated in each of these strategies, the environmental sustainability might be accomplished [34]. Nevertheless, environment is not the only component of sustainability, hence energy and economic components need to be integrally incorporated in any assessment of the selected CCU option, as depicted in Fig. 3.5.

Three considered approaches could interplay an interrelated role in the closure of the carbon cycle: (1) long-lasting products immobilizing CO_2, by developing physical and chemical utilization processes; (2) improvements in process efficiency; (3) deployment of decarbonized energy, including electricity and fuels. In response to these approaches, three CU lines were suggested: carbonation of industrial wastes, RE-driven CO_2 hydrogenation, and CO_2 electrochemical reduction [35]. Additionally, to these three approaches, it has been recommended to the petrochemical industry to address or include strategies leading to the electrification of utilities and processes, switching to renewable (green) H_2; upgrading by-products to chemicals; and incorporating CCSU practices. Regardless of the sustainability efforts, they are

Fig. 3.5 Sustainability components of CCU

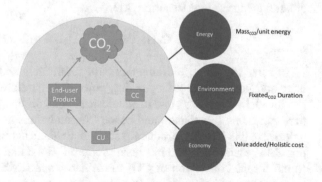

expecting fresh carbon still entering the material cycle, indicating the needs for circularity and for involving the chemical industry into compensating through increased recycling routes. The approach towards a circular industry should ensure that fresh carbon would be provided by a renewable resource, specifically atmospheric CO_2, starting with biomass and transitioning later to CCU [36]. A contribution to the decarbonization of refinery operations has been reported, in which an integrated program targeted to mitigate CO_2 emissions while improving profitability by saving energy [37]. The reduction on emissions could be achieved by process integration, energy management, and energy efficiency improvements. Emissions reductions in the range of 8–12% by process unit, and of 3–6% refinery-wide could be achieved.

3.2.1 Bioeconomy: Role of Renewable Resources

Carbon, one of the most abundant elements on the planet can sustain a thriving economy if used and managed with scarcity criteria. Renewable resources feedstocks such as biomass and CO_2 can be used for the production of biogas, renewable natural gas (RNG), renewable fuels, green hydrogen (H_2), chemicals, and bio-products [38–43]. These biological and renewable resources could support the concept of a bioeconomy [44] that includes industrial and municipal solid wastes (MSW) as feedstocks [45]. So far, developed technologies embedded within the concept of the bioeconomy are well aligned with that of a CE though most of the process plant concepts lack of sustainability and in some instances are CO_2 emitters. These latter technologies could benefit from CO_2 utilization or reuse to close the carbon loop of the Bioeconomy and reach circularity, coming closer to achieve sustainability.

Bioethanol is the most widely used biofuel. While pure ethanol can be regarded as nontoxic and biodegradable, commercialization of bioethanol requires the addition of chemical denaturants, many of which are toxic. Once blended into gasoline, the finished fuel is mainly characterized by the general properties of that fuel. Additionally, while burning bioethanol is considered to be carbon neutral, its production involves CO_2 production, as well. Sugars fermentation is the main reaction used in bioethanol production and in this case, a mole of CO_2 is produced for each mole of ethanol produced (see Reaction 3.1).

$$C_6H_{12}O_6 \rightarrow 2C_2H_5OH + 2CO_2 \qquad (3.1)$$

In some instances, biorefineries burn fossil fuels and by-products, to generate heat and power to input operating units and therefore more additional CO_2 is produced. By-products yield in cellulosic ethanol production is larger than that from first-generation bioethanol, and hence CO_2 emissions are also larger. The need for valorizing this CO_2 by producing fuels and chemicals has been pointed out and a possible general roadmap has been proposed. Good and promising pathways were found for the conversion of CO_2 to methanol and DME [46]. Clearly, biorefining appears then, as non-sustainable and non-circular yet.

Circularity has emerged as a way of managing natural resources, from the linear take-make-use-dispose model into net-zero waste scheme, to preserve biosphere integrity [47]. A net-zero emissions goal has been defined by the European Union (EU) to be achieved by 2050, within the European Green Deal. An example comes from the plastics industry, for which the European Commission has recommended the adoption of biodegradable, bio-based, or both bio-plastics-plastics. Replacing plastics with bio-plastics is not a straightforward task, free of risks, challenges, and/or issues. Some changes in policies, education, R&D and in general, culture and customs are required [48]. In 2018, an updated Agenda revised and redefined policy priorities, adjusting the CE Action Plan into three tiers to support the Bioeconomy Strategy [49, 50]:

1. Strengthen and scale up the bio-based sectors, unlock investments and markets
2. Deploy local bioeconomies rapidly across the whole of Europe
3. Understand the ecological boundaries of the bioeconomy.

There are learnings underneath the principles of the EU CE Action Plan, which include the prevalent role of the bioeconomy in the CE implementation and the great societal challenges involved. The bio-based industries and markets are considered instrumental in demonstration, implementation, and commercialization of products and processes to feed the CE and push it beyond the simple waste management role [51]. The social challenges might also require the bioeconomy (and also CE) to be high on the political agendas. Probably, the first target to be achieved is the combination, or better saying, integration of the two concepts of the bioeconomy and CE, into a **circular bioeconomy** (CBE), therefore emphasizing an increased circularity in material cycles of renewable resources. A comparative analysis of 22 Finnish and Swedish forest biorefineries indicated the lack of sustainability, mainly caused by gaps in closing the material loops. The observed disconnection between academic work and biorefinery endeavors suggests that better communications and education are needed to increase investments and reach CBE implementation [52]. CBE requires a redesign of products, systems, and relationships, without a one solution fits all. Instead, each organization, neighborhood, city, county, and/or country will have a different path to follow. Nonetheless, all players need to be involved and point towards the same horizon. In order to place the bioeconomy in the context of the CE, a clear understanding of what a concerted move towards CE implies for the bio-based business needs to be placed on every stakeholder's plate [53]. Consequently, R&D and education play significant roles, in pursuing these objectives.

In the attempt for pursuing a clean growth, guaranteeing resource security, and reducing fossil fuels dependency, the incorporation of organic waste in the bioeconomy was suggested. The role of the government in policies definition and implementation, as well as in the creation of interaction mechanisms was also established. An integral system and interdepartmental approaches were encouraged, together with vehicles for maximizing industrial synergies [54].

The circularization of the bioeconomy requires the contribution of RE technologies to satisfy future energy demands and reduce greenhouse gas (GHG) emissions. An LCA has been applied to 100 years of electricity generation of solar PV, hydro, and wind to evaluate the impact of their incorporation in a CE, for two different scenarios. REs produced modest savings for an "improved recycling rate" scenario, due to the already low GHGs emissions of this scenario under conventional energy generation. The second "eco-design" scenario did render significant savings in abiotic resource depletion, and therefore eco-design would represent a better climate change mitigation. The main objective of eco-designs is the reduction in resource consumption and waste generation. These designs imply new manufacturing processes and products with extended lifecycle, lower maintenance needs, more reusability, recyclability, or remanufacturability [55].

The other side of the equation is materials and products management, which should include energy optimization and zero waste. In the case of energy optimization, minimization of the environmental impact of biomass processing could be achieved, by maximizing biomass components utilization and effectiveness of any solvent used. Thus, microalgal biomass extraction has been simplified by increasing the solvent usefulness. For this purpose, a switchable-hydrophilicity solvent was employed for a complete extraction of hydrophobic and hydrophilic molecules, in the solvent's opposite-hydrophilicity states [56]. On the zero waste side, bio-based materials could play an important role, providing environmental benefits that include not only waste reduction, but also reducing GHG emissions and harmful chemicals/pollutants volumes, encouraging investment in rural areas and activities, promoting ecosystems and biodiversity conservation, and favoring transition to the circular economy [57]. Naturally and abundantly available materials, such as cellulose and starch are currently being used for paper, packaging, serving objects, bags, and biofuels. However, an increased production and utilization of bio-derived materials and products (e.g., polylactic acid—PLA) should be implemented for a fast transition to a circular economy.

The current linear supply chain followed by society (production → consumption → disposal) governs a two-fold environmental crisis, characterized by resource scarcity and waste overload. Both sides of the coin need to be managed, resource and waste. An increasing relation between resource demand and waste production has been indicated to getting close to a violation of planetary boundaries and of human rights. In this regard, a transition towards a CE would require changes in mentality, industry practices, and policies and regulations in the waste and resource management landscape [58]. MSW remains a vast unexploited source of carbon (and hydrogen), which could be used to produce fuels, chemicals, products, and materials. While waste handling, segregation, classification, and fractionation are advancing quickly, chemical processes and technologies for waste conversion are not being developed, at the same pace. Economic and environmental benefits to produce fuels have been evaluated [59]. Similarly, a waste-to-urea (WtU) technology was also demonstrated to be economically valuable and environmentally advantageous [60].

As expected, a CE pursues a better use of resources and energy, constituting thus the basis for a sustainable future. Nature powers the CBE. In this new economic model, the use of renewable natural capital is emphasized while focusing on the minimization of waste. A CBE targets the production of functional replacement of non-renewable, fossil-based products by environmentally and economically sustainable renewable materials and products [61]. In this context, CBE is a particular case of eco-efficient processes that fixes the CE principles. Both concepts have similar targets, and they somehow overlap, though neither is fully part of the other nor embedded in the other. CBE has also been defined as the intersection of bioeconomy and CE. However, it has been emphasized that the bioeconomy should not be misunderstood as simply a part of the CE and therefore, its agenda and strategy should include differentiating additional and specific topics [62].

3.2.2 Paths to Sustainability

The concept of a CE as mentioned in the Introduction seems to implicitly embed sustainability. However, environmental, and social sustainability is only part of the equation, the other part must include economy and energy. The CE is a systemic approach in which environment, industry/businesses, and society, all together benefit from the economic development. Thus, a sustainable circular economy results in a multi-dimensional system pursuing the best environment quality, human welfare, and pecuniary progress, for current and future generations. All these cannot be attained through just the CE, it has to be a sustainable circular economy (SCE) to achieve multi-dimensional goals [63]. An economic model, based on environmental factors (extended Mankiw-Romer-Weil model) indicated that economic growth was determined by resource productivity, environmental employment, recycling rate, and environmental innovation, within a SCE [64].

3.2.2.1 Approaches and Strategic Measures

An analysis of 45 CE-defined strategies and more than 100 case studies of CE implementation cases indicated that while certain parts of the CE value chain as recovery/recycling and consumption/use are prominently featured, others, such as manufacturing, and distribution are rarely considered. Regarding implementation, many market-ready solutions are already available and so far, the focus has been concentrated on selected products, materials, and sectors. Databases were created for documenting these strategies and the implementation case studies [2].

The economic analysis made by the European Union and conducted by the 27 countries of the Union, prior to defining a strategy for the implementation of a circular economy was based on a model of economic growth, involving the human capital, MSW recycling rate, resource productivity, and green energy use. Factors like resource productivity, labor employed in environmental protection, MSW recycling rate, and RE use were found to be determinants for a CE model [65]. These findings prompt the European Commission to define a new action plan on CE aligned with European Green Deal objectives, including legislative and non-legislative measures for the circularity in regions and cities [66]. The EU has understood that any climate policy and CE are and have to be complementary and mutually reinforcing. The CE implementation cannot be taken as a traditional "product standards" policy. The interconnection among policies and R&D is essential for thriving CE through establishing:

- a framework that systematically addresses trade-offs between the circular and bioeconomy, and between material efficiency and energy use,
- a mechanism to steer and monitor progress, with clear targets and milestones, and
- research, develop and scale up products and processes, circularly integrated.

Additional policies should address the technology push and market pull of CE products and processes, underpinned by new sets of regulations, and embedded within new or adapted business models. Given examples included carbon contracts for difference, sustainable financing, projects with carbon budgets, charges on consumption, tax exemptions and taxes, product standards, and public procurement. A robust and transparent carbon accounting system needs to be developed to monitor the achievements of the CE [67]. Regarding metrics, the recommendation is the creation and definition of indicators to measure both, the what (strategy) and the how (implementing approach). The fact that CE involves the responsible and cyclical use of resources for sustainable development, the measurement of the effects on environmental, social, or economic dimensions should be pursued through indicators' framework. Thus, assessing CE strategies and approaches should involve a set of indicators rather than a single indicator [5]. The set of ten indicators defined by the EU has received criticisms due to their focus on production and consumption, waste management, secondary raw materials, competitiveness, and innovation but lacking on the assessment of (GHG) emissions, the consideration of the whole value chain and the circularity of the approach. Another criticism is on the analysis of the measured indicators since it has failed in identifying data gaps, understanding trends among stakeholders, and helping to anticipate transition barriers [68].

A waste-free chemical industry can be achieved through circular chemistry. An approach of expanding the scope of sustainability to the entire lifecycle of chemical products has been recommended for the optimization of resource efficiency across chemical value chains [69]. However, investment on R&D is required to move the chemical industry, from there to implementing closed loops of materials and products. New concepts are being examined as is the case of a system coupling the Sabatier reaction (R. 34) with the carbon management devices, for long-duration manned space missions. Catalyst development for the Sabatier reaction step was reported in Ref. [70], in which the concept of the integrated system is also described. The heat of the Sabatier reaction is used to heat an air stream that regulates reactor temperature and contributes to the energy balance of the CO_2 management system.

The strategy pursuing the connection between the CE with energy transition found a critical gap in the energy transition framework. This gap concerns the open loop in the fossil fuels lifecycle, particularly regarding the non-energy use of these fuels. The suggested approach recommends the extension of the energy transition discourse to incorporate the transition of non-energy use and the accomplishment of a closed loop of non-energy use. Articulation between CE and energy transition will bring about reductions in energy demand and attain additional GHG mitigation potential [71].

The application of CE to approach environmental restoration, social benefits creation, and clean growth needs to engage stakeholders from academia, government, and industry, such as it has been defined and pursued by the Resource Recovery from Waste program in the United Kingdom. The first principle of the program is maximizing materials value by long-lasting circulation through the economy. Once more R&D and policies articulation is the basis of the approach. Academia-industry collaborative R&D and concerted definition of key themes, regulatory instruments, and stable policies should be embedded within government plans [72].

Aligned with the CE and focused on the minimization of waste, losses, pollution and extraction of virgin resources, strategies targeting the efficient management of resources and utilities could lead to environmental sustainability. This strategy considers a management framework to articulate R&D and deployment (RD&D) of clean technologies, and process modeling and monitoring, for the mitigation of GHG emissions and air and water pollution. A study suggested a disconnection between RD&D, real-life cases, and policies, as well as the lack of economically feasible proven prototypes of the proposed technological solutions [73]. Another approach within this strategy is the resource recovery from waste (RRfW). Typically, within this RRfW approach, a single focus operates, mostly infrastructure investment is almost exclusively focused on energy recovery from waste. RRfW cannot function as standalone, nor short-term approach. It has to be incorporated as a complementary measure to reuse, refurbishment, and recovery of value. Standardization is a need on every single aspect and steps towards circularization, and these standards should include economic, social, and technical metrics [74]. The UK has realized that implementation of RRfW requires investments in the material recovery infrastructure, filling existing gaps on supporting data, and creating an organization or agency for collecting resource stewardship evidences [75].

Energy savings and pollution reduction within the context of a sustainable economic growth was a subject for discussion, in an effort to incorporate contributions from process integration, modeling, and optimization. Relevant approaches included process Integration for sustainable development, energy saving and clean technologies, sustainable processing and production, renewable and high-efficiency utility systems, footprint minimization and mitigation, operations and supply chain management, and waste minimization, processing, and management [76]. Research on controlling carbon emissions from industrial processes has been carried out aligned with strategies of energy efficiency and low-carbon technologies. The major approaches followed in most of the research works involve the integration of process technologies either inward or outward process stages. Approaches for inward process stage consider process integration techniques and efficiency improvements on energy and matter. Meanwhile, CCUS incorporation is the most widely applied basis for outward process stage [77].

3.2.2.2   Business Models

Every and any of the approaches for climate change mitigation places the "business as usual" operations in the unbearable category. The current practices can no longer guide the business interests, but new interests have to be aligned with global societal interests. As pointed out throughout the discussion in the previous chapters and sections, the economic sustainability of a decarbonized world calls for the definition of new business cases, the implementation of new businesses, and the redefinition of business models. The CE, as any other economy is embedded within business models, and a sustainable CE relies on the sustainability of such models. Business models operating in a linear value chain will not necessarily be sustainable when getting cyclic. Although the concept of the CE might be taken as a cyclic integration of linearly integrated businesses, this is an invalid simplification. In fact, CE complexity could be derived from a holistic integration of the concepts graphically expressed in Figs. 3.1 and 3.2. Additionally, CE implementation is underpinned by an orchestrated coordination among business models, policies, and financing organizations [78]. Whether current business models can be adapted or new models need to be created remains a challenge to overcome. Eco-industrial parks have been suggested as potential business models for CE [34], in the search for the business model to become both eco-efficient and eco-effective [79]. The economy within a cyclic model is subjected to multi-dimensional challenges. Since multi-dimensional challenges need multi-dimensional solutions, a new redistribution from economic to social and environmental values through the preservation of the technical value of products and materials is needed. Valid arguments consider the rethinking of costs and values, particularly holistic costs and resource value (see slide 11 in Ref. [79]) in terms of economic, social, technical, and environmental challenges, needs, and problems. The business model must target to become both eco-efficient and eco-effective, and based on transformative technology, by redesigning the production and consumption system to lower "multi-dimensional" costs and optimize all values [79]. This context includes waste management as part of a strategy for consistently improving the quality of waste streams and the value of produced secondary resources, leading to an increase in reprocessing rates and thus on eco-efficiency. However, implicit in this cycle is a call to manufacturers to collaborate by designing materials, components, and products (MCPs) with properties that aid recovery, refurbishing, repair, and recycling to facilitate the closure of the lifecycle. In general, on one side manufacturers often bear little or no direct responsibility for the fate of the used materials and components and for the products behind the factory gates. On the other side is the social perception of a poorer quality of recycled MCPs. A work tries to define a typology to distinguish avoidable and unavoidable designed and created characteristics at all stages of MCPs lifecycle [80].

A view in which the Internet of Things and Industry 4.0 technologies are incorporated into eco-designs has been proposed as enabler for manufacturing companies to establish their CBMs. An empirical validation of the resulting competitiveness of the CBM for eco-design tools and digital technologies in the manufacturing industry has been provided showing fulfillment of sustainability paradigms [81]. Materials effi-ciency is one of the new paradigms of the CE. The volume reduction and the recovery of resources throughout the production and consumption processes in close loops are key for materials efficiency. A business case addressing resource efficiency and avoiding unsustainable consumption brings into consideration industrial symbiosis, governmental interventions, and the transformation of company culture. The design of CBMs needs to consider and use properly: (i) incorporation of biodegradable materials; (ii) modular design for product life extension; and (iii) upcycling for new production processes [82]. Hydrothermal processing of high-density plastics has been proposed as a technology to increase resource recovery in the circularity of the plastic economy. Nearly 100% yields of monomers and high-value compounds, suit-able for chemicals and fuels applications, were obtained. Nine different plastics were tested, including polycarbonate, styrene-butadiene, PLA, poly(ethylene terephtha-late), and poly(butylene terephthalate) [83]. The recycling of durable hydrocarbons has been demonstrated to be far superior to biodegradation. Nonetheless, innovations in catalyst and process technology to enable the thermal degradation/liquefaction and quantitative recovery of low molar mass hydrocarbon oil and gas products are needed. Further, the recyclable hydrocarbon materials meeting the needs of sustainable development would be tailored and integrated into CE [84].

Logistic issues, such as network and geographical distances within a resourcing partnership, were explored to limit C-emissions while promoting resource efficiency and business development. The industrial symbiosis strategy promotes waste-to-resource innovation among other objectives. Learnings from this study include guide-lines for identifying resource partners within a 75 miles' radius, demonstrating the potential of industrial symbiosis for contributing to the transition towards the circular economy [85].

Efforts in the circularization of mineralization include changes in the business model to improve operating costs. The use of mineral rocks as feedstock renders thermodynamically stable products for a permanent and safe sequestration. Carbon-ation reaction rate can be improved by increasing the temperature and pressure, and by decreasing particle size of the mineral. Additionally, solvents that facilitate the extraction of the active elements or components also favor the increase in the reac-tion rate. Besides the reaction rate, another barrier for industrialization is logistics, i.e., transport of materials and fate of products. Transporting the CO_2 to the feed-stock mine seems to be preferred. As mentioned in the previous chapter (Sect. 2.4. End-products and end-users) products are typically used in reclamation and building applications. In the suggested new business model, the use of alkaline solid wastes as an alternative feedstock for calcium or magnesium was acknowledged though further improvements in reaction rates and energy consumption of the process are mandatory [86].

The oil and gas sector within the high emitting industries will be significantly impacted by global decarbonizing strategies. Nevertheless, a study based on the North Sea region has identified specific benefits and drivers when the decarbonizing plan is approached through an integrated circular concept. These include reduction in decommissioning costs and increase in whole lifecycle values of offshore energy infrastructure, a new decommissioning industry involving regional economic development and jobs, contribution to lower carbon economy and energy transition, as well as to environmental restoration, investment into a change from an increasingly demanding oil and gas industries to become "energy companies", and for changing public opinion by supporting low-carbon materials production and energy generation [87].

Critical materials required by RE and low-C energy generation, main pillars of global decarbonization would benefit from circularization and more particularly from the materials efficiency and resource management practices. Babbitt et al. identified the lack of research and industry and policy readiness for the adoption of such practices [88]. In this regard, the technical and/or economic lifecycle of the existing energy infrastructure is predetermined by the lifetime of certain components and more particularly that of the shortest lifetime. Therefore, residual lifetime of components or infrastructure with longer lifetime is typically wasted. Modularization has been suggested as a contributor to the sustainability of the CE. However, the connections between CE and modularization in energy infrastructure remain unknown. The reduction of construction waste and achievement of the closed-loop materials cycle were identified as attention areas [89]. Nevertheless, whatever route to transition into CE is taken, all stakeholders must be in the same page without ambiguity, uncertainties, or misunderstandings. A clarification for concepts such as thoughts to drive policy interventions, CE as a systemic transition of global industrial systems; and the eco-effectiveness role in upgrading business-centered approaches to sustainability is provided by Borrello et al. [90].

Some characteristics specific to Circular Business Models (CBMs) have been identified: multi-stakeholder and multi-company collaboration; value transfer from economy sphere to human development; and resource efficiency and clean technology by organizational learning and technology transfer. A collaboration model was suggested for multi-national companies implementing CE in developing countries, with particular emphasis on the role of local business in knowledge and technology transfer [91]. The long-term sustainability of any CBM is of primary concern when designing the model. The socio-cultural transformations seem to be moving parallelly to the transformation of the industrial systems. New concepts of human behavior and participation have been suggested to discover and unveil productive potential. The concepts of product service systems and "prosumption" were explored in this context to give social intervention in product and services designs [92]. Besides materials efficiency and materials management practices incorporation in CBMs, the new CBMs must consider: (i) investments in resource and energy-efficient processes, (ii) prioritizing energy saving, (iii) close loops for critical and non-renewable materials, and (iv) incorporate CCS and/or C-management technologies [93].

References

1. General union environment action programme to 2020 'living well, within the limits of our planet', in *1386/2013/EU*. The 7th Environment Action Programme: European Union (2013), p. 30. https://eur-lex.europa.eu/legal-content/EN/TXT/?uri=CELEX:32013D1386
2. Y. Kalmykova et al., Circular economy—from review of theories and practices to development of implementation tools. Resour. Conserv. Recycl. **135**, 190–201 (2018). https://doi.org/10.1016/j.resconrec.2017.10.034
3. A.G. Olabi, Circular economy and renewable energy. Energy **181**, 450–454 (2019). https://doi.org/10.1016/j.energy.2019.05.196
4. M.M. Ramirez-Corredores et al., Identification of opportunities for integrating chemical processes for carbon (dioxide) utilization to nuclear power plants. Renew. Sustain. Energy Rev. **150**, 111450, 15 (2021). https://doi.org/10.1016/j.rser.2021.111450
5. G. Moraga et al., Circular economy indicators: what do they measure? Resour. Conserv. Recycl. **146**, 452–461 (2019). https://doi.org/10.1016/j.resconrec.2019.03.045
6. C. Lohan, T. Kylä-Harakka-Ruonala, EESC opinion: Monitoring framework for the circular economy. 2018/C 367/19 Report (The European Economic and Social Committee, Brussels, Belgium, 2018), p. 6
7. Ellen Macarthur Foundation, *Circular economy butterfly diagram* (2019). https://ellenmacarthurfoundation.org/circular-economy-diagram. Accessed July 2021
8. A.P.M. Velenturf et al., Circular economy and the matter of integrated resources. Sci. Total Environ. **689**, 963–969 (2019). https://doi.org/10.1016/j.scitotenv.2019.06.449
9. UN World Commission, *Environment and Development: Our Common Future United Nations* (Oslo, Norway, 1987), p. 300. http://www.un-documents.net/our-common-future.pdf
10. United Nations, *2030 Agenda for Sustainable Development* (Brussels, Belgium, 2015), p. 35. https://www.un.org/ga/search/view_doc.asp?symbol=A/RES/70/1&Lang=E
11. International Advisory Council, *Global Bioeconomy Summit. Innovation in the Global Bioeconomy* (2018), p. 108. https://gbs2020.net/wp-content/uploads/2021/10/GBS_2018_Report_web.pdf. Accessed Feb 2020
12. European Commission, *Innovating for Sustainable Growth—A Bioeconomy for Europe* (European Union, Brussels, Belgium, 2012), p. 64. http://ec.europa.eu/research/bioeconomy/pdf/bioeconomycommunicationstrategy_b5_brochure_web.pdf
13. C. Bauer et al., *New energy externalities developments for sustainability: Technical data, costs, and life cycle inventories of advanced fossil power generation systems*. Deliverable no 7.2—RS 1a, Final Report (Paul Scherrer Institut & Inst. für Energiewirtschaft & Rationelle Energieanwendung, Univ. Stuttgart, 2008), p. 265
14. D. Gielen et al., *Prospects for CO_2 Capture and Storage* (International Energy Agency (IEA), 2004), p. 252
15. R. Guerrero-Lemus, J.M. Martínez-Duart, in *Renewable energies and CO2: Cost analysis, environmental impacts and technological trends-2012 edition*, ed. by ed. R. Guerrero-Lemus, J.M. Martnez-Duart, Lecture Notes in Energy, vol. 3 (2013), p. 395. https://doi.org/10.1007/978-1-4471-4385-7
16. Trade and Agriculture Directorate and United Nations Environment Programme, *Biofuel performance with respect to environmental and other criteria, in economic assessment of biofuel support policies* (2007 MCM mandate). TAD/CA(2008)6/FINAL Report (OECD, IEA, 2008), pp. 36–52
17. P. Moriarty, D. Honnery, What is the global potential for renewable energy? Renew. Sustain. Energy Rev. **16**(1), 244–252 (2012). https://doi.org/10.1016/j.rser.2011.07.151
18. Covestro, *Pure facts—why the circular economy matters* (2016). https://www.covestro.com/-/media/covestro/country-sites/global/documents/k-2016/7_pur_pure_facts_circular_economy_engl.pdf?la=en&hash=CAC6CB692A27D5C4902873B62ED0AC17FF396F6A. Accessed Jul
19. M. North, Across the board: Michael North on carbon dioxide biorefinery. Chemsuschem **12**(8), 1763–1765 (2019). https://doi.org/10.1002/cssc.201900676

20. W.Y. Cheah et al., Biorefineries of carbon dioxide: from carbon capture and storage (CCS) to bioenergies production. Biores. Technol. **215**, 346–356 (2016). https://doi.org/10.1016/j.biortech.2016.04.019
21. P. Styring et al., *Carbon Dioxide Utilisation: Closing the Carbon Cycle* (Amsterdam, Elsevier, 2015), p. 336
22. P. Feron, Growing interest in CO2 -capture from air. Greenh. Gases: Sci. Technol. **9**(1), 3–5 (2019). https://doi.org/10.1002/ghg.1850
23. M. Aresta, Carbon dioxide utilization: the way to the circular economy. Greenh. Gas Sci Technol. **9**, 610–612 (2019). https://doi.org/10.1002/ghg
24. M. Aresta et al., in *An Economy Based on Carbon Dioxide and Water, Potential of Large Scale Carbon Dioxide Utilization*, ed. by M. Aresta (Switzerland AG, Springer, 2019), p. 436. https://doi.org/10.1007/978-3-030-15868-2
25. N.V. Von Der Assen et al., Environmental assessment of CO_2 capture and utilisation, in *Carbon Dioxide Utilisation: Closing the Carbon Cycle*, ed. by P. Styring et al. (Elsevier, Amsterdam, 2015), pp. 45–56
26. N. Wang et al., The circular economy and carbon footprint: a systematic accounting for typical coal-fuelled power industrial parks. J. Clean. Prod. **229**, 1262–1273 (2019). https://doi.org/10.1016/j.jclepro.2019.05.064
27. T.T.D. Cruz et al., Life cycle assessment of carbon capture and storage/utilization: from current state to future research directions and opportunities. Int. J. Greenh. Gas Control. **108**, 103309, 13 (2021). https://doi.org/10.1016/j.ijggc.2021.103309
28. M. Finkbeiner, V. Bach, Life cycle assessment of decarbonization options—towards scientifically robust carbon neutrality. Int. J. Life Cycle Assess. **26**(4), 635–639 (2021). https://doi.org/10.1007/s11367-021-01902-4
29. J.L. Galvez-Martos et al., Environmental assessment of aqueous alkaline absorption of carbon dioxide and its use to produce a construction material. Resour. Conserv. Recycl. **107**, 129–141 (2016). https://doi.org/10.1016/j.resconrec.2015.12.008
30. M. Moreno-Gonzalez et al., Carbon-neutral fuels and chemicals: economic analysis of renewable syngas pathways via CO_2 electrolysis. Energy Convers. Manag. **244**, 114452, 17 (2021). https://doi.org/10.1016/j.enconman.2021.114452
31. National Academies of Sciences-Engineering-Medicine, Gaseous carbon waste resources, in *Gaseous Carbon Waste Streams Utilization: Status and Research Needs* (The National Academies Press, Washington, DC, 2019), pp. 27–38. https://doi.org/10.17226/25232
32. National Academies of Sciences-Engineering-Medicine, in *Negative Emissions Technologies and Reliable Sequestration: A Research Agenda* (The National Academies Press, Washington, DC, 2019), p. 510. https://doi.org/10.17226/25259
33. P.D. Jensen et al., Highlighting the need to embed circular economy in low carbon infrastructure decommissioning: the case of offshore wind. Sustain. Prod. Consum. **24**, 266–280 (2020). https://doi.org/10.1016/j.spc.2020.07.012
34. P.-C. Chiang, S.-Y. Pan, Prospective and perspective, in *Carbon Dioxide Mineralization and Utilization*, ed. by P.-C. Chiang, S.-Y. Pan (Springer, 2017), pp. 441–446
35. E.A. Quadrelli et al., Potential CO_2 utilisation contributions to a more carbon-sober future: a 2050 vision, in *Carbon Dioxide Utilisation: Closing the Carbon Cycle*, 1st edn. (Elsevier Inc, 2014), pp. 285–302. https://doi.org/10.1016/B978-0-444-62746-9.00016-5
36. J.P. Lange, Towards circular carbo-chemicals—the metamorphosis of petrochemicals. Energy Environ. Sci. **14**(8), 4358–4376 (2021). https://doi.org/10.1039/D1EE00532D
37. T. Taraphdar, Reducing Carbon Footprint. Pet. Technol. Q. **16**(2), 1000399, 9 (2011)
38. S.K. Hoekman et al., CO2 recycling by reaction with renewably-generated hydrogen. Int. J. Greenh. Gas Control **4**(1), 44–50 (2010). https://doi.org/10.1016/j.ijggc.2009.09.012
39. W.M. Budzianowski, Negative carbon intensity of renewable energy technologies involving biomass or carbon dioxide as inputs. Renew. Sustain. Energy Rev. **16**(9), 6507–6521 (2012). https://doi.org/10.1016/j.rser.2012.08.016
40. M. Götz et al., Renewable power-to-gas: a technological and economic review. Renew. Energy **85**, 1371–1390 (2016). https://doi.org/10.1016/j.renene.2015.07.066

41. F. Graf et al., Injection of biogas, SNG and hydrogen into the gas grid. GWF, Gas/Erdgas **2**(1), 30–40 (2011)

42. M. Kaltschmitt et al., *Energie aus biomasse: Grundlagen, techniken und verfahren* (Springer, Germany, 2009), p. 216

43. M. Sterner, *Bioenergy and renewable power methane in integrated 100% renewable energy systems: Limiting global warming by transforming energy systems.* Doktor der Ingenieurwissenschaften Thesis from Kassel University Press GmbH, Erneuerbare Energien und Energieeffizienz (Renewable Energies and Energy Efficiency), 23 Sep 2009, p. 230

44. K. Mccormick, N. Kautto, The bioeconomy in europe: an overview. Sustainability (Switzerland) **5**(6), 2589–2608 (2013). https://doi.org/10.3390/su5062589

45. M. Adamowicz, Bioeconomy–concept, application and perspectives. Zagadnienia Ekonomiki Rolnej **1**(350), 29–49 (2017). https://doi.org/10.5604/00441600.1232987

46. G. Centi, S. Perathoner, Catalytic transformation of CO2 to fuels and chemicals, with reference to biorefineries, in *The Role of Catalysis for the Sustainable Production of Bio-fuels and Bio-chemicals*, ed. by K.S. Triantafyllidis, et al. (Elsevier, Amsterdam, 2013), pp. 529–555. https://doi.org/10.1016/B978-0-444-56330-9.00016-4

47. A.P.M. Velenturf et al., A new perspective on a global circular economy, in *RSC Green Chemistry*, ed. by L.E. Macaskie et al. (Royal Society of Chemistry, 2020), pp.3–22. https://doi.org/10.1039/9781788016353-00001

48. A. Di Bartolo et al., A review of bioplastics and their adoption in the circular economy. Polymers. **13**(8), 26 (2021). https://doi.org/10.3390/polym13081229

49. Directorate-General for Research and Innovation, *A Sustainable Bioeconomy for Europe: Strengthening the Connection Between Economy, Society and the Environment* (European Commission, Brussels, Belgium, 2018), p. 107. https://knowledge4policy.ec.europa.eu/pub lication/sustainable-bioeconomy-europe-strengthening-connection-between-economy-societ y_en

50. European Economic and Social Committee, Opinion on a sustainable bioeconomy for Europe: strengthening the connection between economy, society and the environment. Off. J. Eur. Union **4** (2019). https://op.europa.eu/en/publication-detail/-/publication/8ebe492b-a79d-11e9-9d01-01aa75ed71a1/language-en/format-PDF/source-222756549

51. D. Carrez, P. Van Leeuwen, *Bioeconomy: Circular by Nature.* The European Files (2016), pp. 34–35

52. A. Temmes, P. Peck, Do forest biorefineries fit with working principles of a circular bioeconomy? A case of Finnish and Swedish initiatives. For. Policy Econ. 101896, 12 (2019). https://doi.org/10.1016/j.forpol.2019.03.013

53. A. Raudaskoski, Placing the bio-economy in the context of the circular economy (Unpublished Presentation) (2017), 14. https://www.ethica.fi/fi/circular-economy/

54. R. Marshall et al., The organic waste gold rush: Optimising resource recovery in the UK bioeconomy. Resource Recovery from Waste (2018), p. 8

55. J. Gallagher et al., Adapting stand-alone renewable energy technologies for the circular economy through eco-design and recycling. J. Ind. Ecol. **23**(1), 133–140 (2019). https://doi.org/10.1111/jiec.12703

56. A. Cicci et al., Circular extraction: an innovative use of switchable solvents for the biomass biorefinery. Green Chem. **20**(17), 3908–3911 (2018). https://doi.org/10.1039/c8gc01731j

57. R. Shogren et al., Plant-based materials and transitioning to a circular economy. Sustain. Prod. Consum. **19**, 194–215 (2019). https://doi.org/10.1016/j.spc.2019.04.007

58. A.P.M. Velenturf, P. Purnell, Resource recovery from waste: restoring the balance between resource scarcity and waste overload. Sustainability (Switzerland) **9**(9), 1603–1619 (2017). https://doi.org/10.3390/su9091603

59. G. Iaquaniello et al., Waste to chemicals for a circular economy. Chem. Eur. J. **24**(46), 11831–11839 (2018). https://doi.org/10.1002/chem.201802903

60. E. Antonetti et al., Waste-to-chemicals for a circular economy: the case of urea production (waste-to-urea). Chemsuschem **10**(5), 912–920 (2017). https://doi.org/10.1002/cssc.201601555

61. *The Circular Bioeconomy: Knowledge Guide.* (Center for International Forestry Research and World Agroforestry, 2021), p. 4. https://www.cifor.org/wp-content/uploads/2021/03/Flyer%20-%20Knowledge%20Guide_Circular%20Bioeconomy-v4.pdf
62. M. Carus, L. Dammer, The "circular bioeconomy"—concepts, opportunities and limitations. Nova pap. **1**, 9 (2018). www.bio-based.eu/nova-papers
63. A.P.M. Velenturf, P. Purnell, What a sustainable circular economy would look like. Conversation (2020), p. 4. https://theconversation.com/what-a-sustainable-circular-economy-would-look-like-133808
64. C.L. Trica et al., Environmental factors and sustainability of the circular economy model at the European union level. Sustainability (Switzerland). **11**(4), 16 (2019). https://doi.org/10.3390/su11041114
65. M. Busu, Adopting circular economy at the European union level and its impact on economic growth. Soc. Sci. **8**(5), 12 (2019). https://doi.org/10.3390/socsci8050159
66. European Commission, *Circular economy action plan. For a cleaner and more competitive Europe* (European Union, Luxemburg, 2020), p. 28. https://op.europa.eu/en/publication-detail/-/publication/45cc30f6-cd57-11ea-adf7-01aa75ed71a1/language-en/format-PDF/source-170854112
67. V. Rizos et al., Circular economy for climate neutrality: setting the priorities for the EU. Policy Brief (4), 11 (2019)
68. European Economic and Social Committee, Monitoring the implementation of the circular economy. Off. J. Eur. Union (10), 6 (2018). https://op.europa.eu/en/publication-detail/-/publication/d3dad2e1-cc55-11e8-9424-01aa75ed71a1/language-en/format-PDF/source-222756728
69. T. Keijer et al., Circular chemistry to enable a circular economy. Nat. Chem. **11**(3), 190–195 (2019). https://doi.org/10.1038/s41557-019-0226-9
70. H.J. Strumpf et al., Sabatier carbon dioxide reduction system for long-duration manned space application, in *21st International Conference on Environmental Systems* (1991) https://doi.org/10.4271/911541
71. W.M. Chen, H. Kim, Circular economy and energy transition: a nexus focusing on the non-energy use of fuels. Energy Environ. **30**(4), 586–600 (2019). https://doi.org/10.1177/0958305X19845759
72. A.P.M. Velenturf et al., Co-producing a vision and approach for the transition towards a circular economy: perspectives from government partners. Sustainability (Switzerland) **10**(5), 20 (2018). https://doi.org/10.3390/su10051401
73. Y.V. Fan et al., Cross-disciplinary approaches towards smart, resilient and sustainable circular economy. J. Clean. Prod. **232**, 1482–1491 (2019). https://doi.org/10.1016/j.jclepro.2019.05.266
74. P. Purnell et al., New governance for circular economy: policy, regulation and market contexts for resource recovery from waste, in *RSC green chemistry*, ed. by L.E. Macaskie et al. (Royal Society of Chemistry, 2020), pp. 395–422. https://doi.org/10.1039/9781788016353-00395
75. A. Velenturf, P. Purnell, Moving beyond waste management towards a circular economy. (Proc. Valuing the Infrastructure of Cities, Regions and Nations, Leeds, UK, 2017), p. 14. https://www.researchgate.net/publication/316580374_Moving_beyond_waste_management_towards_a_circular_economy
76. J.J. Klemeš et al., Process integration and circular economy for renewable and sustainable energy systems. Renew. Sustain. Energy Rev. **116**, 109435, 7 (2019). https://doi.org/10.1016/j.rser.2019.109435
77. R.K. Sinha, N.D. Chaturvedi, A review on carbon emission reduction in industries and planning emission limits. Renew. Sustain. Energy Rev. **114**, 109304, 14 (2019). https://doi.org/10.1016/j.rser.2019.109304
78. B.K. Sovacool et al., Decarbonizing the food and beverages industry: a critical and systematic review of developments, sociotechnical systems and policy options. Renew. Sustain. Energy Rev. **143** (2021). https://doi.org/10.1016/j.rser.2021.110856
79. A. Velenturf, How to make a business case for resource recovery? in *Proceedings of the Meteor Conference* (Newcastle, UK, 2018), p. 14. https://rrfw.org.uk/2018/09/04/how-to-make-a-business-case-for-resource-recovery/

80. E. Iacovidou et al., Quality of resources: a typology for supporting transitions towards resource efficiency using the single-use plastic bottle as an example. Sci. Total Environ. **647**, 441–448 (2019). https://doi.org/10.1016/j.scitotenv.2018.07.344

81. F.E. Garcia-Muiña et al., Identifying the equilibrium point between sustainability goals and circular economy practices in an industry 4.0 manufacturing context using eco-design. Soc. Sci. **8**(8), 22 (2019). https://doi.org/10.3390/socsci8080241

82. C.-W. Chen, Improving circular economy business models: opportunities for business and innovation. Johnson Matthey Technol. Rev. **64**(1), 48–58 (2020)

83. T. Helmer Pedersen, F. Conti, Improving the circular economy via hydrothermal processing of high-density waste plastics. Waste Manage. **68**, 24–31 (2017). https://doi.org/10.1016/j.wasman.2017.06.002

84. T. Hees, et al. Tailoring hydrocarbon polymers and all-hydrocarbon composites for circular economy. Macromol. Rapid Commun. **40**(1), 1800608, 18 (2019). https://doi.org/10.1002/marc.201800608

85. A.P.M. Velenturf, Initiating resource partnerships for industrial symbiosis. Reg. Stud. Reg. Sci. **4**(1), 117–124 (2017). https://doi.org/10.1080/21681376.2017.1328285

86. W.J.J. Huijgen, R.N.J. Comans, Carbon dioxide sequestration by mineral carbonation: Literature review. ECN-C--03-016 Report (Energy Research Centre of the Netherlands, 2003), p. 53

87. A. Velenturf, Circular economy, the energy transition and sustainable design for decommissioning in North sea oil and gas, in *Proceedings of the Offshore Decommissioning Conference* (Virtual, UK host., 2020), p. 13

88. A.P.M. Velenturf et al., Reducing material criticality through circular business models: challenges in renewable energy. One Earth **4**(3), 350–352 (2021). https://doi.org/10.1016/j.oneear.2021.02.016

89. B. Mignacca et al., Modularisation as enabler of circular economy in energy infrastructure. Energy Policy **139**, 111371, 11 (2020). https://doi.org/10.1016/j.enpol.2020.111371

90. M. Borrello et al., Three propositions to unify circular economy research: a review. Sustainability (Switzerland). **12**(10), 22 (2020). https://doi.org/10.3390/SU12104069

91. J.L. Mishra et al., Collaboration as an enabler for circular economy: a case study of a developing country. Manag. Decis. **59**(8), 1784–1800 (2019). https://doi.org/10.1108/MD-10-2018-1111

92. K. Hobson, Closing the loop or squaring the circle? Locating generative spaces for the circular economy. Prog. Hum. Geogr. **40**, 88–104 (2016)

93. R. Lantto et al., Going beyond a circular economy: a vision of a sustainable economy in which material, value and information are integrated and circulate together. Ind. Biotechnol. **15**(1), 12–19 (2019). https://doi.org/10.1089/ind.2019.29156.rla

Chapter 4
Perspectives and Future Views

4.1 Introduction

Although governments from industrialized countries have committed to the COP 26 goals, an analysis of the achieved progress shows very little advances. The position of ten C-intense/energy-intense economic sectors within the decarbonization transition phase [1] was incorporated in a conventional S-shaped curve for the penetration of low-carbon technologies [2], showing how early each of them was advancing through the transition and indicating how far they are from the COP 26 goals. Globally the transformation of the Energy Systems needs to be significatively accelerated, shifting away from fossil fuels but also improving efficiency, electrification, and new fuels, targeting to meet net-zero commitments. More than ever the immediate future needs that energy gets generated from any available alternate (low-carbon) source, preferably renewable sources [3] but also including more nuclear into the energy mix.

Previous chapters have placed the focus of attention to CU as a vehicle for value creation and as an enabler for a circular economy, which all together if sustainably integrated would inevitably lead to the net-zero goal of less than 1.5 °C global temperature increase. Nevertheless, the emitted CO_2 needs firstly to be captured, for its subsequent utilization. Hence, the crucial role of CC in meeting climate change mitigation goals is unquestionable. However, a 90% capture rate seems to be everybody's assumption regardless of whether or not veracious, or whether a higher limit is achievable. It has been found that economic feasibility would depend on whether capture rates above 98% become achievable, which in turn depends on the CO_2 concentration of the flue gas or gas stream to be treated [4]. Additionally, the currently existing scale gap between the emitted mass (Gton/y) and what finds application or use in industry (Mton/y) is of 3–4 orders of magnitude. Therefore, capture rates >98% are required not only to overcome economic barriers but also to ensure environmental appealing. Additionally, CO_2 recovery from the capturing stage is energy-intense. Therefore, and as pointed out in Fig. 2.3, intensification needs call for CC–CU integrated systems, and in general terms, integration of reactions and separation processes might be a valid approach for overcoming this matter [5, 6].

Previously we have discussed and described a variety of reactions involving CO_2 occurring both in nature and most desirably in chemical plants. Table 4.1 collects these reactions, notice that reactions R.1 through R.4 were described in Chap. 1, as reactions 1.1 through 1.4; reactions R5 through R42 were discussed in Chap. 2, as reactions 2.1 through 2.38; and reaction R43 was considered in Chap. 3 as reaction 3.1. As mentioned above, very few have reached commercial scale and all are far from being performed with optimal energy efficiency, nor providing the best benefits on C-intensity or C-footprint. Thus, needs, challenges and gaps still exist and will be discussed in the next sections.

4.2 Challenges, Needs and Gaps

A research gap represents an unsolved part of some area of knowledge that has been thoroughly studied with the concomitant consequence that technology development cannot be completed unless one bridges the identified gap. On the other hand, a research challenge concerns a much broader concept, which is related with studies that have not been carried out by the scientific community yet, or studies which are in the very preliminary stages. Usually, the identification of research challenges and the fulfillment thereof will result in research gaps. A gap analysis of any sort of process requires the assessment of the current state of the system under evaluation and the identification of the ideal future state. Such analysis must identify possibilities of improvement, i.e., the opportunities. Based on such analysis, a plan must be built aiming at bridging the gap. It must be borne in mind that technologies evolve fast, hence the gap analysis must consider the concept of innovation, or rather, gap analysis must not just rely on existing technologies.

Although the scientific basis underpinning, explaining, and understanding the causes and effects of climate change and advances are progressing rapidly, there persist unanswered questions. As pointed out above, climate change was originally hypothesized and since then, a body of scientific evidence has been built, evaluated, and seriously debated to derive into today's knowledge and explanations. Nevertheless, advancing the science of climate change is a standing challenge for preparing the responses and developing the required technologies for mitigating its causes. The US National Research Council has identified seven research areas in connection to this challenge, which includes a comprehensive climate observing system, improved climate models and other analytical tools, investments in human capital, and linking research and decision making through partnerships with action-oriented programs [7].

This Section summarizes the findings from the previous chapters and identifies needs, challenges and gaps, which will require R&D for addressing, assessment, overcoming or solving those and advancing the different technologies into the market plate. These technologies are meant to add value, understanding value-added as the

Table 4.1 Reactions involving CO_2 discussed in previous chapters

Reaction topic	R#	Reaction	Quoted remark
CO_2 cycle	R1	$CO_2 + CO_3^{2-} + H_2O \rightarrow 2HCO_3^-$	
	R2	$6CO_2 + 6H_2O + photons \rightarrow C_6H_{12}O_6 + 6O_2$ $C_6H_{12}O6 + 6O_2 \rightarrow 6CO_2 + 6H_2O + heat$	
	R3	$CO_2 + CaSiO_3 \rightarrow CaCO_3 + SiO_2$	
	R4	$CO_2 + CaCO_3 + H_2O \rightarrow Ca^{2+} + 2HCO_3^-$	
Carbonation	R5	$2H^+ + H_2O + (Ca, Mg, Fe)SiO_3 \rightarrow Ca^{2+}, Mg^{2+}, Fe^{2+} + H_4SiO_4$	
	R6	$H_2CO_3 + Me^{2+} \rightarrow MeCO_3 + 2H^+$	
	R7	$Mg_2SiO_4 + 2CO_2 \rightarrow 2MgCO_3 + SiO_2$	$\Delta H_{298\,K} = -89$ kJ/mol
	R8	$Mg_3Si_2O_5(OH)_4 + 3CO_2 \rightarrow 3MgCO_3 + 2SiO_2 + 2H_2O$	$\Delta H_{298\,K} = -64$ kJ/mol
	R9	$Ca_2SiO_4 + 2CO_2 \rightarrow 2CaCO_3 + SiO_2$	$\Delta H_{298\,K} = -90$ kJ/mol
	R10	$CO_2 + NH_3 + H_2O \rightarrow (NH_4)HCO_3$	
	R11	$2NH_3 + CO_2 \rightleftharpoons H_2N\text{-}COONH_4$	$\Delta H_{298\,K} = -117$ kJ/mol
	R12	$H_2N\text{-}COONH_4 \rightleftharpoons (NH_2)_2CO + H_2O$	$\Delta H_{298\,K} = 15.5$ kJ/mol
Syngas	R13	$CH_4 + H_2O \rightarrow CO + 3H_2$	$\Delta H_{298\,K} = 206$ kJ/mol
	R14	$CO + H_2O \rightarrow CO_2 + H_2$	$\Delta H_{298\,K} = -41.15$ kJ/mol
	R15	$CO_2 + CH_4 \rightarrow 2CO + 2H_2$	$\Delta H_{298\,K} = 247$ kJ/mol
	R16	$CO_2 + H_2 \rightarrow CO + H_2O$	$\Delta H_{298\,K} = 41.15$ kJ/mol
	R17	$CH_4 + \frac{1}{2}O_2 \rightarrow CO + 2H_2$	$\Delta H_{298\,K} = -36$ kJ/mol
	R18	$CH_4 + 2O_2 \rightarrow CO_2 + 2H_2O$	$\Delta H_{298\,K} = -802$ kJ/mol
	R19	$CH_4 \rightarrow C + 2H_2$	$\Delta H_{298\,K} = 74.85$ kJ/mol
	R20	$2CO \rightarrow C + CO_2$	$\Delta H_{298\,K} = -172$ kJ/mol
Chemical looping	R21	$CH_4 (CO, H_2) + MeO \rightarrow CO_2 + H_2O (CO_2, H_2O) + Me$	
	R22	$Me + \frac{1}{2}O_2 \rightarrow MeO$	
Iron loop	R23	$Fe_2O_3 + 3CO \rightarrow 2Fe + 3CO_2$	
	R24	$3Fe + 4H_2O \rightarrow Fe_3O_4 + 4H_2$	
	R25	$4Fe_3O_4 + O_2 \rightarrow 6Fe_2O_3$	

(continued)

Table 4.1 (continued)

Reaction topic	R#	Reaction	Quoted remark
MDR-CL	R26	$4MeO + CH_4 \rightarrow$ $4Me + CO_2 + 2H_2O$	Methane reduction
	R27	$2Me + 2CO_2 \rightarrow 2MeO + 2CO$	CO_2 reforming
	R28	$2Me + 2H_2O \rightarrow 2MeO + 2H_2$	Steam reforming
	R29	$CH_4 + CO_2 \rightarrow 2CO + 2H_2$	Net reaction (R. 22 + R. 23 + R. 24)
	R30	$2Me + O_2 \rightarrow 4MeO$	Air oxidation
SEB	R31	$CaO + CO_2 \leftrightarrow CaCO_3$	$\Delta H_{298\ K} = -178$ kJ/mol
	R32	$CaO + 2CO \leftrightarrow CaCO_3 + C$	$\Delta H_{298\ K} = -350$ kJ/mol
Hydrogenation	R33	$CO_2 + 3H_2 \rightarrow CH_3OH + H_2O$	$\Delta H_{298\ K} = -11.8$ kcal/mol
	R34	$CO_2 + 4H_2 \rightarrow CH_4 + 2H_2O$	$\Delta H_{298\ K} = -165$ kJ/mol
	R35	$CO + 2H_2 \rightleftharpoons CH_3OH$	$\Delta H_{298\ K} = -91$ kJ/mol
Carboxylation	R36	$CO_2 + 2ROH \rightarrow RO\text{-}COO\text{-}R + H_2O$	
	R37	$CO_2 + e_{aq} \rightarrow COO^-$	
	R38	$RH + OH \rightarrow R \cdot + H_2O$	
	R39	$COO^- + R \rightarrow RCOO^-$	
Industrial abatement	R40	$CaO + NH_4X + H_2O \rightarrow$ $CaX_2 + NH_4OH$	
	R41	$NH_4OH + CO_2 \rightarrow$ $(NH_4)_2CO_3 + H_2O$	
	R42	$(NH_4)_2CO_3 + CaX_2 \rightarrow$ $CaCO_3 + 2NH_4X$	
Bioeconomy	R43	$C_6H_{12}O_6 \rightarrow 2C_2H_5OH + 2CO_2$	

features, characteristics or attributes present in final products or services, which provide different types of benefits to the end-user. The value-added can be estimated as the difference between the price of a product or service and the cost of producing it [8]. In fact, for a given industry or company, the value-added corresponds to the difference between its total financial income and its total expenses with inputs such as materials, labor, supply, and services. In that sense, considering that CO_2 is an abundant raw material, one might admit that, in principle, reactions transforming CO_2 into other molecules would be an easy way of generating value-added products. However, this is not always the case. The cost of carbon capture must not be over-looked in the productive chain. Such cost may be somewhat high, depending on the CO_2 occurrence. Verticalization is often considered as an efficient tool to add value to commodities. However, this is not always true, since we are currently facing a huge business-process revolution in the world; hence, the concept of de-verticalization is being proposed [9]. Additionally, the high costs have been associated with worsening the public perception of CCS that has created further barriers in its deployment.

Surprisingly, rather than addressing the cost challenges and problems, the definition and execution of new policies have been proposed [10].

Besides costs, CC and particularly DAC face enormous technical challenges. Regardless of these technical limitations and the current poor economics, Hanna et al. calls for political support for an urgent deployment of DAC [11]. Firstly, this urgent deployment will face the lack of commercially available technologies to handle the massive amounts due to removal. In fact, the IAE is currently reporting in the Web announcement of the DAC report [12] that globally there are 19 operating plants capturing about 10,000 $tonCO_2$/y, which represents a gap of seven orders of magnitude between emissions (Gton) and capturing (~0.5 kton/plant). Scale-up is in progress and units promising capturing rates of 1 Mton/y are available for licensing though no commercial demonstration has been reported [13]. However, CC is not the main focus of this book and discussing these challenges fall out of our scope, more details on the current status of scaling and availability are given by McQueen et al. [14]. Nonetheless, there is a scientific challenge that is worth pointing out here. The fact that there is a minimum CO_2 atmospheric concentration that could trigger off photosynthetic activity is worth emphasizing once more. The reader is invited to revisit the discussion (and references therein) included in Sect. 1.2 of Chap. 1. The lack of peer reviewed papers documenting this finding and the lack of systematic studies are indications that further research is needed and the scientific challenge still prevails. Hence, whether there is a threshold concentration of CO_2 that could stop photosynthesis needs to be undoubtedly determined.

In Table 4.1, the examples of reactions involved in the uses and applications of CO_2 have been summarized. These reactions lead either to value-added products or to render a service or application that provides a benefit to the end-user, therefore collectively they regard to CO_2 valorization. Carbon dioxide valorization is a vehicle for creating economic benefits for producing from fuels to bulk and commodity chemicals (e.g., inorganic materials) and even to specialty products with biological activity such as pharmaceuticals and functionalized polymeric materials while carbon is fixated during a prolonged time (see examples in Fig. 4.1).

New chemistries that either use or convert CO_2 into building blocks or final products are required. So far, numerous catalysts have been tested for promoting the studied reactions in Table 4.1, most of them existing formulations that were designed for the particular reaction but probably for a different feedstock or vice versa. More specificity is needed for the catalyst to be designed for these reactions with higher activity, selectivity, and stability (including lower deactivation). The manufacture of these efficient catalysts could benefit from using cheap and naturally abundant compounds, and the of use recyclable materials.

The concept of eco-design and its incorporation in the CE was discussed in Sect. 3.2.1 of Chap. 3. Its implementation needs are evidently pointed out in that discussion. However, persuading industry and the manufacturing sector to adapt their practices and develop new eco-designed processes and products remains a challenge. A proposal to include regarding policies in the EU Circular Economy Package and the related legislation has been documented [15].

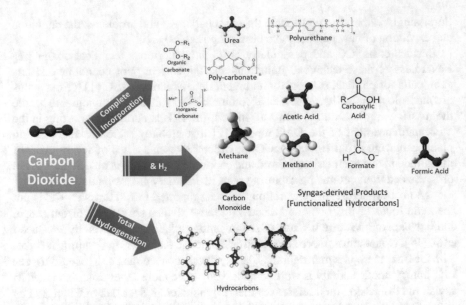

Fig. 4.1 Examples of value-added products rendered from the conversion of CO_2

Energy generation, the main responsible for the CO_2 emissions carries also the greatest of all scientific and engineering challenges, i.e., developing, implementing, and deploying net-zero new energy systems to satisfy the global energy demand that drives economic growth sustainably. Several needs of the sector to overcome this challenge have been identified and include a price for GHG emissions, the definition and evaluation of associated costs for financial, technical and market risks of the new technologies, determination of existing gaps in infrastructure and supply chain, creation of incentives and last but not least more and higher quality information [16]. Some of these needs can be satisfied by a new set of better policies and regulations, but others require supporting R&D and even participation of the Education and Media Sectors. Regarding CU, the thermodynamic stability of CO_2 implies high activation energy for CO_2 reactions, desirably to form energy-rich intermediate streams or products with longer lifecycle. Most of the studied processes result in high energy intensity, therefore creating the need for the development of more energy efficient processes and/or for devoting efforts in energy intensification of the considered pathways.

4.2.1 Challenges and Needs

Inorganic valorization of CO_2 concerns reactions that use CO_2 as feedstock and minerals or inorganic compounds as co-reactants (reactions R. **1**, R. **3**–R. 12) aiming at producing derivatives which present an interesting added value. However, as far as value-added is concerned, an important issue that should not be disregarded is the production of critical minerals. Indeed, mineral production will play a critical role in the energy transition to a cleaner world [17]. The term, critical minerals is used here to refer to elements, compounds and minerals playing a vital role in energy security (availability, accessibility, and affordability). It must be borne in mind that clean energy technologies require more minerals than the traditional fossil-based traditional technologies. Thus, as energy transition gets closer and accelerates, the magnitude of the (critical) minerals in the energy sector will also increase, making it the leading consumer of these materials. IEA forecasted the share of clean energy technologies sector in the demand for exemplary critical elements, for two different scenarios, the Stated Policies Scenario (STEPS), and the Sustainable Development Scenario (SDS). STEPS provided indications of where the energy system was heading, based on a sector-by-sector analysis of current policies and policy announcements, while SDS refers to what would be required to meet Paris Agreement goals. Other examples include, for instance, typical electric cars that require six times the mineral inputs than those necessary to produce a conventional car. Moreover, an onshore wind plant requests nine times more mineral resources than a usual gas-fired plant. The new battery-powered technologies require important amounts of cobalt, graphite, lithium, nickel, and manganese. Rare earth elements are critical constituents of permanent magnets that are essential for wind turbines and EV motors. A tremendous amount of copper is necessary to build electricity networks, as well as aluminum. Regarding the catalytic transformation of CO_2 into hydrocarbons or chemicals, some of the catalysts require cobalt, ruthenium, nickel, and other critical elements in minor amounts. Surely, the increase in mineral consumption will depend on the technology deployed [17].

As previously highlighted in Chap. 2 (Sect. 2.1), carbonation comprises the processes of transformation of CO_2 into alkaline or alkaline earth carbonates, bicarbonates, or carbamates. These are typical mineralization processes which imitate natural processes. Generally, carbonation may be divided into two main areas. The first one concerns "MC" or "mineral sequestration", which is the fixation of CO_2 to yield inorganic carbonates while the second one regards the development of processes which use CO_2 as a feedstock aiming at producing value-added chemicals.

The main advantage of such processes is the fact that they are thermodynamically spontaneous. In fact, the carbonation reaction is exothermic and can theoretically produce energy. Nevertheless, some energy-consuming steps are needed aiming at improving the process efficiency. In fact, some mineral pretreatment, which may comprise several steps such as mechanical separation, crushing, milling, and grinding are often required before the reaction process [18]. Among these, important energy-consuming process steps are, for instance, the grinding of the mineral raw material to improve its reactivity and the compression of the CO_2 feed. Therefore, commercial processes require much energy to increase conversion by preparing the solid reactants. Furthermore, the use of additives is often needed. Such additives must be regenerated and recycled, which requires more energy.

On the other hand, although the reaction thermodynamics is favorable, the same does not occur when kinetics is concerned. Carbonation reactions are very slow.

Obviously, reaction kinetics can be accelerated by raising the reaction temperature. Nonetheless, mainly when reactions in aqueous phase are concerned, high temperatures may enhance the formation of gaseous CO_2, thereby hampering carbonates' precipitation. The challenges of improving the kinetics and the energy efficiency of the Abo Akademi route for producing $Mg(OH)_2$ from serpentinite mineral were thought to be achieved using a pressurized fluidized bed, which in turn introduces new challenges such as the fluidization at high pressures. Although reaching equilibrium in 10–15 min represents an improvement in the kinetics, there was no report on the effect on energy efficiency. These results were considered to be promising enough to pursue scaling up studies [19]. The search for energy efficiency improvements moves to ideas on process integration and intensification. The exothermicity of the reaction was suggested to drive an autothermal reactor, run under hydrostatic pressurization in a vertical plug flow design for an underground installation. The process kinetics, conversion and energy balances were the basis for mathematical modeling, which was then applied to define particle size, solids loading, pumping rate, and reactor dimensions, to guarantee autothermal operation. The identified conditions were considered to maximize carbonation efficiency and enable recoverable heat generation, to optimize energy efficiency [20]. However, no validation of model prediction was presented. For all these reasons, the technology is still in the development stage, being not yet ready for implementation.

According to IPCC Special Report on Carbon dioxide Capture and Storage, Chap. 7 [21], "*The best case studied so far is the wet carbonation of the natural silicate olivine, which costs between 50 and 100 US$/ton$_{CO2}$ stored and translates into a 30–50% energy penalty on the original power plant. When accounting for the 10–40% energy penalty in the capture plant as well, a full CCS system with MC would need 60–180% more energy than a power plant with equivalent output without CCS.*" It is clear that MC costs are much higher than those of other sequestration routes [22]. Hence, such technology can only be commercially acceptable should reaction kinetics be improved.

An interesting point that concerns cost reduction is the potential use of inorganic industrial wastes. Depending on the type of wastes, the reaction kinetics may be improved. However, factors such as the site where both the CO_2 source and the industrial wastes are produced must be taken into account. In fact, logistics plays a major role in this case. Furthermore, a more detailed mapping of mineral resources such as silicates must be carried out. Moreover, for these sources to be used, a thorough study of environmental constraints is required [23–25].

The potential direct use of CO_2 or as a feedstock in processes to produce value-added chemicals is an important topic in the field of CCUS. However, depending on the life cycle assessment of the chemical process deployed, the amount of CO_2 sequestered and stored will significantly vary. Also, depending on how long the product will remain in use, the lifetime of such sequestered CO_2 will oscillate. At any rate, products from captured CO_2 are supposed to have a rather long lifetime before CO_2 is released by any type of process. A life cycle assessment of the global process of MC comprising the steps of mining, size reduction process, waste disposal, and site restoration has been performed. Results indicated calculated additional annual

Table 4.2 Lifetime of CO_2 carbonation products

Product	CO_2 consumption ($Mton_{CO2}/Mton_{product}$)	Lifetime (y)
Urea	65	0.5
Methanol	<8	0.5
Inorganic carbonates	3	$10-10^2$
Organic carbonates	0.2	$10-10^2$
Polyurethanes	<10	$10-10^2$
Food	8	$10^{-2}-10$

CO_2 emissions of 0.05 $ton_{CO2}/ton_{sequestered-CO2}$ [26]. The cost was estimated to be about 14 US\$/ton_{CO2} stored, being the capital cost of about 20% of the total.

The extension of the lifetime of the products rendered from the conversion of CO_2 is one of the challenges of the valorization pathways. The Special Report of Mazzotti et al. [21] collects the extended life-span of some carbonation, chemical products and materials obtained from CO_2, ranging from days to centuries (products considered in this book are included in Table 4.2). Clearly, the need for holistic LCA could help in defining the potential cost-benefits of each of these products.

The evaluation of the data depicted in Table 4.2 clearly indicates that the amount of CO_2 that might be sequestered via chemical process is in fact negligible compared with total anthropogenic carbon emissions. Hence, the deployment of CO_2-based industrial chemical processes as a way of mitigating climate change does not have a brilliant future since the scale is too small, the storage times are too short, and the energy balance is rather unfavorable. Nevertheless, much research is still being carried out aiming at discovering new chemical routes which will take advantage of the unique chemical characteristics of CO_2 as a potential raw material for the polymer industry [27].

The abundant variety of derived products that could be obtained from the CO_2 organic valorization, specially fuels and chemicals gives rise to significant opportunities. Nevertheless, it is important to be aware of the challenges involved in the development of these processes to attain sustainability in terms of energy and economy, e.g., energy efficiency improvements and profitability. Regarding the net-zero requirements on the reduction of CO_2 emissions by 2030, the IEA highlights the influence of the commercially available technologies. However, the significant reductions required by 2050 depend on technologies, from which almost half of them are currently in the demonstration or prototype phase [28]. Therefore, the most prominent R&D challenge, strongly emphasized here, is the need of devoting as much effort as possible to bring these new technologies to implementation.

Concerning CO_2 transformations, the thermodynamic stability of CO_2 compels for the use of high temperatures to carry out the conversion reactions. Additionally, most of the CO_2 reactions are endothermic. Since currently the required heat is produced by the combustion of fossil sources, these chemical processes result in high energy and carbon intensity. Consequently, the need for process intensification are clearly derived.

The main MDR thermodynamic challenge is its high thermodynamic potential, $\Delta G°_{298} = 174.6$ kJmol^{-1}, which requires the use of high temperatures to diminish the critical free energy. A common problem in CO_2 conversion reactions at high temperatures, including MDR, is catalyst deactivation, particularly by coke deposition. Since endothermic reactions are favored at high temperatures, thermodynamically coke formation is facilitated [29]. However, combination of MDR with the exothermic MPO (R. 17) and/or total oxidation reactions (R. 18) could be used for energy intensification [30–32]. The use of biogas as feedstock for MDR shows great potential as a greener alternative.

Even though syngas could also be produced by combining the electrochemical reduction of CO_2 with the electrolysis of water [33] this process is not a well-established technology. As described in Chap. 2, Sect. 2.3.1.3, two types of processes have been studied and developed in the co-electrolysis of CO_2 and water referred as LTE (T < 100 °C) and HTE (T $\leq$ 600 °C). At the HTE conditions, water is present in the form of steam, for this reason, co-electrolysis of CO_2 and steam is a representative form of HTE [34]. A combination of chemical and electrochemical processes, for the total reduction of CO_2 in high-temperature electrolysis (HTEs) is a potential application of a hybrid type of processes [35]. Challenges are related to improve materials integrity and stability under HTE conditions to achieve longer life cycle applications. These concerns together with the lack of solutions might preclude the community from overcoming the challenges and from advancing the technology into larger scale processes. In general, besides these technical reasons, there might be also some other probable economic challenges to overcome to increase the number of industrial HTE operations [36] that historically have been very limited. One of the cost factors negatively impacting electrocatalytic applications is the need for a pure or highly concentrated CO_2 stream. Therefore, additional steps for product separation and purification, and for recovery and recycling of unconverted reactants are needed, and will also negatively impact the process economy [36–38].

The thermochemical cycles, including the chemical looping concept (described in Chap. 2, Sect. 2.3.1.2) may offer a high efficiency option for reducing costs and energy demands from the power and chemical sectors. High pressures have been found to negatively affect C-capturing though it would be effective in process intensification of the chemical looping process. The lack or low experience on building and operating large-scale CLC systems has created barriers for the commercial implementation of this technology [39]. Besides accumulating experience, collecting more data and information for generating more knowledge is needed in areas such as oxygen carrier durability, technical feasibility of downstream high-temperature valves and filters, before moving forward to overcome the scale-up challenges. TEA results indicated that the reactor temperature creates constraints for applications on combined cycle power generation [40]. Therefore, this area calls for materials optimization (in terms of cost-effectiveness) as well as for the development of materials management practices and guidelines.

The hydrogenation of CO_2 is considered an important strategy for CO_2 recycling rather than releasing it to the atmosphere [41–43]. Among CO_2 hydrogenation products, methanol and hydrocarbons can be used as fuels in internal combustion engines without or with minimum modification of existing logistic infrastructure for storage, distribution, and retailing [44, 45].

Besides its fuel application, methanol is also an intermediate or raw material for petrochemicals manufacture and it is considered an energy carrier for a clean and sustainable energy future. Commercially available process technologies mentioned in Chap. 2, Sect. 2.3.2.1 (Olah, Lurgi and CAMERE) have demonstrated the technical feasibility of the catalytic hydrogenation of CO_2, for methanol production. Although the Olah process is currently in operation in Iceland, all these processes can be considered as emerging technologies. Main drawbacks of these processes are their low conversion and poor selectivity, due to the occurrence of multiple side reactions. Conventional methanol synthesis catalysts have been and are being used for this application. Therefore, the development of new catalyst formulations showing higher activity and selectivity is required to increase methanol productivity. Energy management is probably the main challenge for these technologies for which integration to power plants has been recommended for economic improvement. The poor technical performance, energy efficiency and economy all together represent a clear need for an integrated energy-chemical system, in which both parts i.e., energy generation and chemical process are cost-effective technologies.

Processes for electrocatalytic hydrogenation reactions might be a potential alternative though these are in their infancy. The limited solubility of CO_2 in aqueous media is a challenge for this application, although in high throughput processes, a wastewater treatment application may not be greatly affected [46]. As per previous catalytic processes, improving FE, current density and over-potential needs the development of new more active and selective electrocatalysts.

Since the CO_2 methanation process (i.e., Sabatier reaction, R. 34) is an exothermic reaction, controlled by thermodynamic equilibrium, the greatest challenge involved in this process is the control of the enormous heat release [47]. When methanation is a part of a PtG system, the reaction heat can be channeled into electric energy generation or otherwise managed [48–51]. When biogas is used as feedstock for the Sabatier reaction, the RNG produced has higher energy value than the original biogas. PtG-Oxycombustion hybridization is an innovative technology which allows the direct comparison between biogas upgrading and flue gas methanation. Operating volatile RE sources (e.g., solar, wind) for chemical production systems requires a deeper understanding of their dynamic operation modes in order to identify control trajectories for a time optimal reactor start-up, thus avoiding distinct hot spot formation [16, 52]. Although biogas seems a convenient raw material for Power-to-Gas (PtG) systems, it must be borne in mind that the cost of H_2 remains a significant factor in determining the cost of renewable methane. In addition to biogas as an alternative feedstock, a mixture of CO and CO_2 resulting from methane reforming has also been studied in methanation. Such mixture requires alternative methanator designs and methanation catalyst. Based on experimental results, new catalytic systems have been proposed to operate at high temperature, high pressure methanation [53].

The two-step mechanism proposed for the carboxylation reactions, involving the RWGS reaction (R. 16) as first step has derived in new pathways for carbonylated products. The reversibility of the WGS reaction (R. 14) and the well documented catalytic details open numerous opportunities for the interconversion of the CO_2/H_2 and the CO/H_2O pairs through the WGS equilibrium [54]. Carboxylation can be undergone with epoxides (R. 36), producing organic carbonates or polycarbonates, with CO_2 resulting a building block, economically and environmentally convenient. Ionizing radiation has also been employed to induce carboxylation production of carboxylic acids, employed to produce various organic derivatives. None of the several materials tested as catalysts has demonstrated acceptable activity, stability and/or recovery through the considered processes, neither have them been tested at relevant industrial-scale (real feedstocks and operating conditions). Polycarbonates production provides an alternative to use massive amounts of CO_2, which opens countless opportunities to initiate R&D programs covering from the most fundamental aspects of science up to scale-up and industrialization activities. The direct use of flue gas as feedstock for this type of reaction represents the main challenge in the search for a cost-effective solution for static emitters. The conceptualization of new process designs with this objective in mind needs to start immediately. Clearly, integration to emitting sources may be considered as an intensification vehicle for the most promising and best cost-effective solution. Nonetheless, a holistic assessment of each potential solution also requires development and standardization of combined TEA/LCA methodologies.

Bioelectrochemical synthesis offers a promising solution to transform organic and inorganic carbon to high value-added products. The reduction of CO_2 into energy-rich species such as C1 or C2+, is considered as an artificial photosynthesis since it converts exhaust carbon into working carbon.

For an industrial exploitation of such idea, several challenges have to be solved, among them an efficient system for solar energy capture, efficient charge separation, physical separation on the catalysts for the oxidation and reduction processes, processing cost reduction for the recovery of energy-rich species by producing water insoluble organics, efficient and selective catalysts for water oxidation and CO_2 reduction, catalyst manufacturing from cheap and naturally abundant compounds, using recyclable materials [55]. The advances in solar energy and concentrated power reactors [56, 57] are closing the knowledge gaps, opening new opportunities for the future. However, more investment and efforts are needed in this area, which could be considered as the basis for a sustainable future [58].

CO_2 incorporation into renewable feedstocks such as biomass, biomass intermediates and bio-products provides pathways to the potential production of materials of interest for low-carbon energy carriers. Co-valorization of CO_2 and by-product glycerol (from the biodiesel industry) can be an example of low-carbon bio-products for applications in pharmaceutical, cosmetics, and plastics industries [59]. However, additional challenges appear from product separation and recovery, particularly when considering the purity required by the utilizing industries.

Ionizing radiation has also been used for inducing reactions of CO_2 with renewable feedstocks and other co-reactants [60]. Irradiation has also been employed to induce mutations and for facilitating deconstruction of lignocellulosic macromolecules, as discussed in Chap. 2. The fact that CO_2 irradiation produces highly reactive species creates numerous opportunities for promoting or inducing a myriad of reactions.

However, the success of using irradiation depends on achieving predictive controls on the propagation and termination reactions. Additionally, these intermediates formed by the ionizing radiation include activated species, as well as radicals, which could be considered for polymerization reactions such as those mentioned above that not only could use a massive amount of carbon but also incorporates the CO_2-moiety in long-life products.

The need of specific catalysts, the development of new and more energy efficient processes and the problem of residues are within challenges, requiring solutions to achieve a better exploitation of the renewable resources and to improve CO_2 fixation.

Regarding end-users and end-products, the nature and composition of the GHG emissions of the high energy- and/or C-intensity sectors differ, their advances in decarbonization and their position within the transition phase are also different, while some commonalities have been identified. For the lowest advanced sectors (e.g., steel, cement, plastics, heavy road transport, aviation, and shipping), the priority challenge to address is the first demonstration, testing and deployment of new technologies, which could be facilitated by governmental supporting policies and incentives. These actions will also benefit those sectors that had started to transition (e.g., power, cars, and buildings) but still need to increase their market share of low-carbon technologies [2]. Concerns about the competitiveness of decarbonizations and/or the effect of decarbonization on the competitiveness of the sector (products, processes, or services) is a challenge to overcome for sectors with high trade rates (e.g., aviation, agriculture). This particular challenge might call for the development of coordinated standards or of carbon pricing. The timing for this efforts depends on the sector, aviation seems to be more urgent while agriculture appears more as long-term [2]. In general, one of the main challenges of the industrial sectors is the development, adoption, implementation, and deployment of net-zero heat technologies. Since the industrial heat requirements is one of the reasons for the "hard-to-decarbonize" qualification of the sector, net-zero heat technologies represent a significant breakthrough for the transition phase of industry [61].

The beginning of the new century found that fundamental chemical science was not involved in the discussion of the technological, economic, and ecological aspects of carbon management regardless of its potential for identifying and addressing the underlying chemical questions. Two extremes were suggested to range the discussion of needs and challenges: (i) avoid CO_2 emissions by using renewable fuels systems and (ii) use CCS while still relying on fossil fuels. In this regard, the identified areas needing R&D included RE storage, generation, and utilization; CO_2 conversion; advanced engine and fuel systems; industrial implementation of CCS and CCU; longer-life C-based renewable products; and effective use of C-containing materials and products [62].

Sustainability of the CE is not one of its intrinsic characteristics, and only an SCE can lead to a sustainable development. Velenturf and Purnell reviewed the published literature on systems ecology and on CE to derive a framework for the design, implementation, and evaluation of an SCE, based on ten principles whose needs and challenges are summarized below [63]:

1. *Beneficial reciprocal flows of resources between nature and society*: the reciprocity of the flows of materials challenges for a synchronization of rates for the resource extraction and the return to the environment, requiring a regenerative and absorptive capacity of the Earth,

2. *Reduce and decouple resource use*: In line with the previous principle, this one challenges for an effective and sustainable materials management,

3. *Design for circularity*: CE requires eco-design whose challenges and needs were discussed above. Primarily, this principle challenges the optimization of stocks and the need for closing material flow loops: the minimization of resource extraction and waste generation, maximization of value creation, and designing for reintegration of materials into natural biogeochemical processes at end-of-life,

4. *Integration of multi-dimensional value by using CBMs*: Industry is challenged to adopt, adapt, or develop innovative business models and governance frameworks to take full advantage and become benefited by circular practices,

5. *Transform consumption*: this principle calls for social and cultural changes in consumerism, implying more sharing, more recycling and even innovative development of products reuse,

6. *Citizen participation in sustainable transitions*: this challenges the development of systems to enable a participatory role of citizens in policy development and decision-making, as well as the cultural change to develop a social willingness to participation,

7. *Coordinated participatory and multi-level change*: the need for high level of coordination is clear when considering the required activities for CE development, integration and implementation, involving all actors and stakeholders from government, industry, civic sector, consumers, and academia,

8. *Multiple circular economy solutions by mobilizing diversity*: the complexity of required changes and of inter-relations among the different industrial sectors challenges for plurality of perspectives and solutions, networks that facilitate knowledge exchange and learning across society,

9. *Political economy for multi-dimensional prosperity*: it challenges the development and implementation of integrated multidimensional systems, such as political-economic systems, connecting, for instance, GDP growth to long-term multi-dimensional prosperity in environmental, social, and economic terms,

10. *Whole system assessment*: CE challenges for a comprehensive and a whole system approach that allows the understanding of challenges and the anticipation of impacts from proposed solutions in a precautionary manner, as well as the monitoring of progress and definition of follow-up guidelines.

Finally, the critical role played by NETs in the pathway to net-zero and achieving the 1.5 °C goals of the Paris Agreement, ratified in COP26 was discussed in Chap. 2 (Sect. 2.4). However, the most promising NETs still carry enormous challenges, which should be overcome to reach that 1.5 °C goal. Their large-scale deployment might represent an economic and environmental burden [64], which needs to be faced and addressed. Implementation difficulties and vulnerability to disturbances were found for low-cost technologies (e.g., A/R, SCS and BC) [65]. Food security might be significantly compromised through BECCS [66] while DACCS, EW and OF might fall into the prohibitively expensive category [13]. The issues found with the carbon accounting methods and systems led to recommend paying more attention to the selection of the accounting method to match the needs of the purpose, and a more holistic examination of the entire system, considering total changes in emissions and removal. However, if the purpose is to construct static descriptions of possible net-zero worlds, attributional methods must be used. Attributional LCA is not an appropriate method to estimate system-wide changes caused by the use of NET. Several distinct carbon accounting challenges were determined, the importance of which varies across different NETs. The use of greenhouse gas accounting methods should be used to estimate system-wide changes to report emissions and removals when they are actually occurring [67].

4.2.2 Gaps

Regardless of existing and potential technologies of CO_2 utilization, previous discussions have made clear that the largest gap concerns the scale difference between emissions (37 Gton/y) and utilization (<250 Mton/y), which is 3–4 orders of magnitude. Hence, processes using massive amounts of CO_2 as feedstock and producing large scale commodities are undoubtedly the most promising ones, to achieve net-zero goals.

Regarding CCUS, capture technologies are quite well-known nowadays though in fact, there are still few knowledge gaps in this area, such as the integration of capture, transport and storage in industrial-scale projects [68]. Moreover, legal and regulatory issues for implementing CCS on an industrial-scale are not fully settled and hinder its implementation. Also, a more detailed (and holistic) LCA of the different CCUS routes has to be carried out on both current and emerging technologies [69].

The need for CCUS intensification led to the most relevant gap that would be the development of one-step integrated CO_2 capture and conversion technology (Fig. 2.3). Although plenty of innovative routes have been disclosed in the literature, such routes are still in the proof-of-concept phase. For instance, hydrogenation of CO_2 captured in amine solutions to formate has been proposed [70]; also, a combined system containing a superbase, a poly (ethylene glycol) (PEG), ionic liquid, and amino acid has been tested to produce in one-step urea, carbonates and formate salts [71], and the conversion of captured CO_2 into syngas [72] or methane [73] has also been studied. One-step integration of CC and conversion will ameliorate energy efficiency and likely will favor the economy[74]. Indeed, this integrated route will reduce costs by eliminating desorption, compression, transportation, and storage steps.

On the CC side, the concept of dual function materials (DFM) has been recently developed to incorporate conversion, besides absorption capabilities. One such materials comprises methanation catalytic sites, (e.g., metallic Ru or Ni) and CO_2

absorbent characteristics (e.g., CaO, MgO) [75]. These new materials exhibited high stability and outstanding catalytic properties, resulting in promising industrial-scale intensified processes.

Methanation of CO_2 has been considered as either the heat absorbing reaction in thermochemical cycles or the heat provider of chemical heat pipes, at the site where energy is generated. Being a gas, methane would be conveniently transported via a pipe system. Gaps in this route include the optimization of operating parameters such as reaction temperature, pressure, and composition. This optimization will depend on the nature of the primary energy source, as well as on the type of end-use for the delivered energy [76]. Moreover, a fully integrated system must be designed to allow its long-range transport using an existing distribution network [77].

Electrolytic processes [78] are trying to address the very high activation energy [79] of CO_2 reduction reactions. Since processes are currently in laboratory or bench scale, the issues concerning scalability must be resolved, including high over-potential, low FE, and current densities. Such problems bring about energy losses and overall processes inefficiencies. Techno-economic studies regarding the viability of a commercial plant are still lacking [80]. Moreover, new catalysts are to be developed to overcome slow kinetics derived from the high activation energy [81–84].

As an example of gaps identification, let us take back the mineralization via carbonation technical and economic challenges: operating costs reduction by reducing solid losses caused by raw material attrition; capacity increase while maintaining solid structural integrity by increasing solid surface area; energy efficiency improvement by reducing heat demand, and optimizing heat management systems; and solids handling improvement. Addressing these challenges would involve R&D including (i) the development of low-cost active raw materials, with thermal and chemical stability, low attrition rates, low heat capacity, high CO_2 absorption capacity, and high CO_2 selectivity, (ii) system configurations of improved energy efficiency and performance. Overcoming the mentioned issues, would successfully lead to reductions in capital and in energy costs, as well as an improved reliability [85].

Broadly speaking, some knowledge gaps and uncertainties may be highlighted. The first one concerns costs. Indeed, little is known about industrial costs of several technologies that have been proposed. Secondly, the potential of distinct markets cannot be overlooked. Market sizes for different technologies can be variable. Some markets can be small but then, should the product yielded via a CO_2 utilization route be cheaper, a new much larger market might arise. Eventually, shifts in government policies in municipal, state and federal levels must be considered as a potential gap to be discussed. Many types of policies such as taxes and mandates are found nowadays in different government levels. More encompassing CO_2 policies are expected to be issued, which will surely affect all aforementioned gaps [86]. During the last years, a decrease in investments in energy efficiency and in renewables represents one of the negative global trends in decarbonization [87]. Therefore, closing the gap in incentivizing policies has a sense of urgency.

Another gap, which must not be disregarded concerning circular economy is related to the use of non-energy fossil raw materials. Although the main use of fossil raw materials is in the generation of energy, they find applications in construction materials, chemical feedstocks, lubricants, solvents, waxes, and other products. Petroleum derivatives used in the production of plastics or natural gas as a feedstock in fertilizers are typical examples. Hence, it is highly recommended that energy transition policies be extended to include non-energy fossil raw materials [88].

Reference [89] provides a panorama of several alternative energy supply technologies, trying to highlight main challenges, gaps and opportunities. In principle, by implementing such technologies, the generation of CO_2 would be reduced. Hence, these are complimentary efforts in the area of decarbonization. However, such technologies also present challenges and gaps. Nuclear power, hydropower, wind power, bio power, solar (PV and concentrated solar power, CSP), geothermal energy, marine hydrokinetic power and fuel cells are good examples of alternative technologies that, in principle, will help reduce CO_2 emissions. The electric power generation system, highly compounded by fossil fuels would benefit by CCUS for transitioning into a cleaner generation of energy. Maturity of electric power generation technologies can leverage global integration policies, minimizing risks and reducing costs. Besides the use of coal or natural gas with CCUS and nuclear power plants, alternative renewable technologies such as wind and solar are growing fast. Also, new technologies are being developed, being fuel cell and marine hydrokinetic power quite promising [85]. Moreover, in a scenario of deep carbon emission reductions, recycling-induced elimination of materially retained carbon releases to the atmosphere, along with the use of recycled materials, show a great potential as alternative methods [90].

Studies on China's [91] and Turkey's [92] scenarios in which power generation, industry and household are the main ones responsible for CO_2 emissions indicated that only stopping the use of coal as raw material [77] would lead to net-zero emissions achievement. Additionally, "Renewables" were found to be unable to meet China's electricity demand, unless feedstock input is supplemented with fossil fuels. Furthermore, land-use, land-use change and forestry are recognized as accessory tools to mitigate CO_2 increasing concentration [91].

Eventually, structural social changes are critical on attaining decarbonization objectives, even R&D targets. Potential actions to bridge such gaps include [93]:

1. Generation of zero-carbon electricity—this gap may be addressed by implementing decarbonization of existing power plants with low-cost technologies and the implementation of IESs may help to reduce carbon emissions. Certain IESs are good examples of circularity, for instance, in cases where thermal energy normally wasted after electricity/power production is subsequently used to perform other energy demanding services e.g., cooling, heating, or humidity control.

2. Replacing energy generated by fossil fuels, by end-use electrification—Decarbonizing the transport sector has to be the main goal of government's policies. Unfortunately, although governments are committed to embracing it, actions are not sufficient to achieve the levels of reductions required. EVs may be a solution though, if it could be facilitated by government policies, including, for instance, fiscal incentives and infrastructure investments. Biofuels/bioblendstocks must

not be disregarded since they provide quite a promising solution for the sector, provided the convenient technologies reach the market. Innovative synthetic renewable fuels are also welcome, e.g., DME.

3. Green synthetic fuels—Green synthetic fuels are energy carriers which may reduce CO_2 emissions. Hydrogen and RNG or SNG may be convenient solutions for the transition to a decarbonized energy system since they can easily be converted into heat or electric power to be used as peak-shaving agents, i.e., agents to level out peaks in the demand of electricity or gas, thereby increasing the flexibility of both electric and gas grids. In addition, implementation of the concept of E2M (Energy to Molecules), for using surplus energy to transform abundant and low-cost molecules such as CO_2 or H_2O in chemical intermediates or final products [94].

4. Smart power grids—capable of using multiple sources and provide energy to various end-users, without losses in efficiency, reliability, and cost.

5. Materials efficiency—the circularity concept, or rather encouraging practices of reuse/recycling, or sustainability in consumption and manufacturing to minimize wastes must be adopted.

6. Optimized sustainable land-use—the agriculture sector must be committed to the minimization of GHG emissions from deforestation, direct and indirect uses of fossil fuels, and deployment of industrial fertilizers and of livestock. The philosophies of Land-Use, Land-Use Change and Reforestation ought to be a continuous exercise to the entire agribusiness.

The research gaps are multiple whenever CCUS is concerned. At any rate, a "cradle to grave" life cycle assessment has to be carefully performed to identify them all, aiming at implementing alternative and intelligent solutions to bridge such gaps.

4.3 Envisioned Future

The abatement of climate change (a humankind obligation to future generations) needs to be undertaken during the first half of the twenty-first century. Our reliance on carbonaceous fuel resources over the past century has elevated economic activity globally. This dependency on fossil resources will continue through the coming decades, causing the generation of large volumes of gaseous carbon waste, especially carbon dioxide. Meanwhile, rapid advancements in carbon mitigation and removal technologies have taken place but their implementation and deployment are needed to reduce detrimental impacts on climate in continued use of fossil fuels. Since CO_2 generation cannot be cut to zero at once, the processing and valorization of several thousands of tons of CO_2 (per day in multiple plants) need to couple with a decarbonizing strategy of the energy supply, by using energy resources with low-carbon emissions. As mentioned above, all UN countries have agreed to execute action plans and implement technologies to overcome the challenges that global warming is imposing and to reach the established plateau of <1.5 °C. This agreement implicitly established that the mitigating strategies are no longer valid and need to be replaced by a total abatement fight. Accordingly, accelerated reductions in the atmospheric CO_2 concentration are required and there is global consensus that more than one strategy and approach will be required, including but not limited to:

1. Accelerated development and implementation of carbon capture (CC), utilization (CU) and storage (CS), CCUS technologies,
2. Zero-carbon electricity,
3. Electrification of end uses, particularly for end-user sectors based on fossil fuel energy,
4. Improvements in energy efficiency of processes for energy generation and consumption,
5. Green synthetic fuels, including the development and deployment of a wide range of synthetic fuels, adapted to carbon-intense sectors,
6. Smart power grids systems are capable of shifting among multiple sources of power generation and various end uses, without detriment on efficiency, reliability, and cost,
7. Materials efficiency, warranting choices and flows to continuous efficiency improvement and minimize waste,
8. Sustainable land-use, involving mainly the agriculture sector, to minimize GHG emissions from deforestation, industrial fertilizers, livestock, and direct and indirect fossil fuel uses,
9. Cultural and social changes.

Nevertheless, whatever the strategy is, energy security cannot be jeopardized and the sense for an urgent implementation requires the commitment of all stakeholders. Based on these statements, a paradox was derived involving two contradictory facts, i.e., one is the difficulties (almost impossibility) to decarbonize fully and fast enough to limit global warming this century to well below 2 °C and the other is that considering what is at stake, such rapid decarbonization is imperative [95].

The IEA recently reported a roadmap to achieve net-zero emissions by 2050, in the Energy Sector, emphasizing the role of the governments in the implementation of energy and climate policies. Also, it highlights the relevance of the urgent commercialization of clean energy technologies, and the innovation opportunities in three key technologies namely, advanced batteries, hydrogen electrolyzers, and direct air capture and storage [28]. Thus, globally "Net-zero" Initiatives are being pursued, implemented, or defined all over and a sense of urgency has been given to it. At this point, the needs for net-zero approaches, involving decarbonized and decarbonizing processes are clearly established. Similarly, the replacement of a linear economy for a circular one is also clear. However, establishing and implementing these process technologies, cyclically integrated, no matter what, does not warrant a better future. The key lies in sustainability.

The transition towards a better future may start by identifying what is reality and what is myth. Smil in his book [96], not only identified some of the myths around the use of various types of energy, but also tried to debunk those and emphasized the fake of the proposed incredibly high rates of the transition from traditional energy sources to RE sources. Within the scope of this present book, the Smil's identified prevalent and persistent myths include: (i) realization or implementation of accelerated developments, (ii) value promise (performance level, hypothetical achievements and potential impact) of new technologies sustainability, (iii) decline of the fossil energy prevalence due to resource scarceness, (iv) return of nuclear energy, (v) broadening deployment of RE and their prevalence in the energy mix, (vi) finding geoengineering solutions to meet the global warming challenges. His advice includes: (i) distrust any strong, unqualified claims regarding accelerations or value promises and the timing for their implementation and achievements, (ii) give the benefit of the doubt to the adaptability of traditional resources and technologies, (iii) keep an open, critic and assertive mind and views for unproven new energies and processes, (iv) be aware and be an informed citizen on the investments and costs needed for the desired changes, and on the time scale for meeting the goals [96].

Although progress has been accomplished on carbon management needs, in the chemical sciences R&D, some of the opportunities for future R&D, identified back in 2001 [62] still prevail for future achievements, e.g., new CO_2-based polymers; cost-effective catalysts for polymerization, hydrogenation, electrochemical and photo-chemical processes; catalysis in supercritical CO_2; proton-based electrochemical and photochemical reduction and hydrogenation. Investigation of these areas not only needs presently acceleration but also results and outputs should be incorporated into technology development programs, in the same accelerated manner, without ignoring their potential for becoming sustainable and for integration with other processes into circular models. The present consideration of accelerating NETs development and/or implementation is critical in achieving the Paris Agreement 1.5 °C goal. A globally broad implementation of NETs in the future requires a suitable monitoring and measuring system for carbon accounting that solves the issues and problems found with current systems. Additionally, system-wide GHG accounting methods should be used to evaluate, monitor and report emissions and removal, as a function of time [67]. A reliable accounting method is the only mechanism to develop credibility on the impact and achievements of NETs and for clearly establishing their role in mitigating climate change. NETs integration into a circular economy relies on an effective application of carbon management and on policies than incentivized it. The industrial sector of the future is one based on sustainable CBMs, which considers the circular integration of decarbonizing processes that produce decarbonized products. In the future, manufacturing and commercial businesses, industry, and society, in general, will avoid the direct use of fossil resources and products, preferring those with a renewable or decarbonized origin. Society will make smart decisions based on sustainability criteria, giving special value to the health of the planet and to the life that it nourishes.

Circularity will play a key role in the integration of decarbonizing (energy efficient) processes, embedded in effective carbon (and materials, in general) management for rendering decarbonized (eco-designed) materials and products. As has been discussed in previous chapters, sustainable and decarbonized businesses are more likely to fit within a circular model than in the conventional linear models. There-fore, a cleaner and more amicable environment, under a mitigated climate change will be brought in a circularly operated economy/businesses future.

4.4 Final Remarks

This book presented perspectives for a future based on an in-depth review of the literature and on the advances in science and technology to control and mitigate GHG-related problems, such as CO_2 abatement through its capturing and transformation. Regarding CCUS, the manipulation of reaction conditions and process parameters, as well as the design and development of highly active catalysts for more energy efficient processes have been emphasized. In this way, the acceleration of the development of decarbonization routes will be enabled, facilitating achievement of the net-zero goals.

Regardless of the efforts linking the anthropogenic CO_2 emissions to climate change, there is an important part of society that still refuses to believe it. There-fore, an important issue to address today for a decarbonized future is to establish as an undoubted fact that there is an anthropogenic influence on the increase in the planet's temperature. This has to be established prior to any attempt to develop appropriate procedures to accomplish a circular economy. An existing paradox seems to indicate that "the larger the development achieved, the higher the probability of influencing climate change", nonetheless assuring the right balance among economic and environmental interests has to be prioritized. Part of society believes that anthro-pogenic emissions are responsible for having brought us here. Therefore, it is high time society takes a more proactive role in making the desired changes to happen. Nevertheless, four potential scenarios [97] could be taken into account when consid-ering a decarbonized future. Two conformed by a proactive society acting either at global or at regional scale, and the other two conformed by a reactive society, which does not believe in climate change that either act collaboratively at global scale or does nothing. Obviously, society or social groups within the latter scenarios will not consider the implementation of decarbonization routes and the proactive part of society will need to take educational actions and/or more drastic measures for the desired future to have any possibility. An example is the Oil & Gas industry, which is one of the most important CO_2 emitters. It is worth mentioning that the world will not admit any fossil fuel-based industry without CCUS should the first two scenarios prevail. Before the coronavirus pandemic, all companies were keen to reduce CO_2 emissions. However, the difficult market conditions (demand drop) have curbed the enthusiasm and the velocity with which decarbonization routes will be implemented. The other side of the coin is a society always fossil fuel hungry, for which non-RE consumption seems impossible, making the energy transition an important and definitive need. This transition will allow a smoothly broader penetra-tion of RE, at the time that fossil fuels from the global energy mix will be decreased, to accomplish the unquestionable necessity of mitigating climate change.

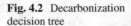

Fig. 4.2 Decarbonization decision tree

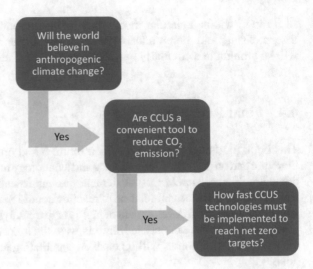

Eventually, the future of decarbonization routes will comprise the line of thinking sketched in Fig. 4.2.

A sustainable circular economy needs to be supported by fundamental principles centered on the protection of the climate and/or in general the environment, while resources are preserved and energy usage is optimized, by maximizing energy efficiency and effectiveness of its uses. Since energy is key in human development, its efficient and effective use is not enough, three other aspects combined into energy security have to come into play namely, availability, accessibility, and affordability. The latter two are currently under severe scrutiny due to the detection of vast regions of energy poverty. Therefore, fighting the root-causes of climate change cannot overlook the energy issues and cannot hinder, nor create barriers for human development. Thus, human development must be looked into with new lenses. Sustainability is based upon the triple bottom line model, in which an equilibrium among "three Ps": profit, people, and the planet are pursued. Sustainability is also intimately related to ethics, in a sense that "ethos", the Greek word for "home" is no longer the house, the city or the country one inhabits. "Ethos" is the entire planet Earth. Should an ethical attitude not be performed, decarbonization will not be a solution.

References

1. D.G. Victor, Deep decarbonization: a realistic way forward on climate change. Yale Environment 360 (2020), p. 8. https://e360.yale.edu/features/deep-decarbonization-a-realistic-way-forward-on-climate-change
2. D.G. Victor et al., Accelerating the low carbon transition: the case for stronger, more targeted and coordinated international action. UK Government Department for Business (2019), p. 71. https://www.brookings.edu/wp-content/uploads/2019/12/Coordinatedactionreport.pdf

3. Mckinsey & Company Global energy perspective—executive summary. USA. (2022), p. 28. https://www.mckinsey.com/~/media/McKinsey/Industries/Oil%20and%20Gas/Our%20Insi ghts/Global%20Energy%20Perspective%202022/Global-Energy-Perspective-2022-Execut ive-Summary.pdf

4. P. Brandl et al., Beyond 90% capture: possible, but at what cost? Int. J. Greenh. Gas Control. **105**(103239), 16 (2021). https://doi.org/10.1016/j.ijggc.2020.103239

5. J.F. Múnera et al., Combined oxidation and reforming of methane to produce pure H2 in a membrane reactor. Chem. Eng. J. **161**(1), 204–211 (2010). https://doi.org/10.1016/j.cej.2010. 04.022

6. Y. Li et al., Oxidative reformings of methane to syngas with steam and CO2 catalyzed by metallic Ni based monolithic catalysts. Catal. Commun. **9**(6), 1040–1044 (2008). https://doi. org/10.1016/j.catcom.2007.10.003

7. National Research Council, *Advancing the Science of Climate Change*. (The National Academies Press, Washington, 2010), p. 526. https://doi.org/10.17226/12782

8. A. Hayes, Value-added (2022), https://www.investopedia.com/terms/v/valueadded.asp. Accessed 22 April

9. A. Raskin, N. Mellquist, The new industrial revolution: de-verticalization on a global scale, (2005), https://www.alliancebernstein.com/cmsobjectabd/pdf/research_whitepaper/r28 453_deverticalization_051215.pdf. Accessed April 2022

10. S. Pianta et al., Carbon capture and storage in the United states: Perceptions, preferences, and lessons for policy. Energy Policy **151**(112149), 8 (2021). https://doi.org/10.1016/j.enpol.2021. 112149

11. R. Hanna et al., Emergency deployment of direct air capture as a response to the climate crisis. Nat. Commun. **12**(1), 368, 13 (2021). https://doi.org/10.1038/s41467-020-20437-0

12. S. Budinis, Direct air capture, report in preparation (2021), https://www.iea.org/reports/direct-air-capture. Accessed March 2022

13. D.W. Keith et al., A process for capturing CO2 from the atmosphere. Joule **2**(8), 1573–1594 (2018). https://doi.org/10.1016/j.joule.2018.05.006

14. N. Mcqueen et al., A review of direct air capture (DAC): scaling up commercial technologies and innovating for the future. Prog. Energy **3**(3), #032001, 23 (2021). https://doi.org/10.1088/ 2516-1083/abf1ce

15. V. Rizos et al., The role of business in the circular economy: Markets, processes and enabling policies. Centre for European Policy Studies. Brussels, Belgium (2018), p. 80. www.ceps.eu

16. M.A. Brown et al., Carbon lock-in: Barriers to deploying climate change mitigation technologies, in *Barriers to Climate Change Mitigation Technologies and Energy Efficiency* (Nova Science Publishers, Inc. 2011), pp. 1–166

17. World Energy Outlook Team, The role of critical minerals in clean energy transitions. IEA. Paris, France (2022), p. 287. https://www.iea.org/reports/the-role-of-critical-minerals-in-clean-energy-transitions/executive-summary

18. W.J.J. Huijgen et al., Energy consumption and net CO_2 sequestration of aqueous mineral carbonation. Ind. Eng. Chem. Res. **45**(26), 9184–9194 (2006). https://doi.org/10.1021/ie0 60636k

19. R. Zevenhoven et al., Carbon storage by mineralisation (CSM): Serpentinite rock carbonation via Mg(oh)2 reaction intermediate without CO_2 pre-separation, in *Proceedings of the 11th International Conference on Greenhouse Gas Control Technologies, GHGT* 2012, vol. 37 (Kyoto, Elsevier Ltd, 2013), pp. 5945–5954. https://doi.org/10.1016/j.egypro.2013.06.521

20. R.M. Santos et al., Integrated mineral carbonation reactor technology for sustainable carbon dioxide sequestration: 'CO2 energy reactor'. in *Proceedings of the 11th International Conference on Greenhouse Gas Control Technologies, GHGT 2012*, vol. 37 (Kyoto, Elsevier Ltd, 2013) pp. 5884–5891. https://doi.org/10.1016/j.egypro.2013.06.513

21. M. Mazzotti et al., Mineral carbonation and industrial uses of carbon dioxide, in Special report on carbon dioxide capture and storage, in *Intergovernmental Panel on Climate Change (IPCC)*, ed. by B. Metz, et al. (Cambridge University Press, UK, 2005), pp. 319–338

22. A.A. Olajire, A review of mineral carbonation technology in sequestration of CO_2. J. Petrol. Sci. Eng. **109**, 364–392 (2013). https://doi.org/10.1016/j.petrol.2013.03.013

23. W.K. O'Connor et al., Carbon dioxide sequestration by direct mineral carbonation: Process mineralogy of feed and products. Miner. Metall. Process. **19**(2), 95–101 (2002). https://doi.org/10.1007/bf03403262

24. W.K. O'Connor et al., Aqueous mineral carbonation. DOE/ARC-TR-04–002 Report. National Energy Technology Laboratory. Albany, Oregon. USA. Mar 15, 2005. 463 pp

25. W.K. O'Connor et al., Carbon dioxide sequestration by ex-situ mineral carbonation. Technology **7**(S), 115–123 (1999)

26. P.S. Newall et al., CO_2 storage as carbonate minerals. PH3/17 Report. IEA GHG; CSMAConsultants Ltd. Cornwall, UK. February 2000. p. 185. https://ieaghg.org/docs/General_Docs/Reports/Ph3_17%20Storage%20as%20carbonates.pdf.

27. S. Kaiser, S. Bringezu, Use of carbon dioxide as raw material to close the carbon cycle for the german chemical and polymer industries. J. Clean. Prod. **271** (2020). https://doi.org/10.1016/j.jclepro.2020.122775

28. International Energy Agency, Net zero by 2050, a roadmap for the global energy sector. IEA. Paris, France. (December 2021), p. 224. https://www.iea.org/reports/net-zero-by-2050

29. M.R. Goldwasser et al., Combined methane reforming in presence of CO2 and O2 over LaFe1-xCoxO3 mixed-oxide perovskites as catalysts precursors. Catal. Today **107–108**, 106–113 (2005). https://doi.org/10.1016/j.cattod.2005.07.073

30. C. Jensen, M.S. Duyar, Thermodynamic analysis of dry reforming of methane for valorization of landfill gas and natural gas. Energy Technol. **9**(7) (2021). https://doi.org/10.1002/ente.202100106

31. J. Hunt et al., Microwave-specific enhancement of the carbon–carbon dioxide (Boudouard) reaction. J. Phys. Chem. C **117**(51), 26871–26880 (2013). https://doi.org/10.1021/jp4076965

32. A.T. Bell, The impact of nanoscience on heterogeneous catalysis. Science **299**(5613), 1688–1691 (2003). https://doi.org/10.1126/science.1083671

33. A. Alcasabas et al., A comparison of different approaches to the conversion of carbon dioxide into useful products: Part I CO2 reduction by electrocatalytic, thermocatalytic and biological routes. Johns. Matthey Technol. Rev. **65**(2), 180–196 (2021). https://doi.org/10.1595/205651321x16081175586719

34. J.E. O'brien et al., High-temperature electrolysis for large-scale hydrogen and syngas production from nuclear energy—summary of system simulation and economic analyses. Int. J. Hydrog. Energy **35**(10), 4808–4819 (2010). https://doi.org/10.1016/j.ijhydene.2009.09.009

35. S. Hernández et al., Syngas production from electrochemical reduction of CO2: current status and prospective implementation. Green Chem. **19**(10), 2326–2346 (2017). https://doi.org/10.1039/C7GC00398F

36. R. Küngas, Review—electrochemical CO2 reduction for CO production: Comparison of low- and high-temperature electrolysis technologies. J. Electrochem. Soc. **167**(4), 044508 (2020). https://doi.org/10.1149/1945-7111/ab7099

37. R. Küngas et al., Systematic lifetime testing of stacks in CO2 electrolysis. ECS Trans. **78**(1), 2895–2905 (2017)

38. J. Artz et al., Sustainable conversion of carbon dioxide: an integrated review of catalysis and life cycle assessment. Chem. Rev. **118**(2), 434–504 (2018). https://doi.org/10.1021/acs.chemrev.7b00435

39. H. Fang et al., Advancements in development of chemical-looping combustion: A review. Int. J. Chem. Eng. **2009**, 710515 (2009). https://doi.org/10.1155/2009/710515

40. M. Osman et al., Review of pressurized chemical looping processes for power generation and chemical production with integrated CO2 capture. Fuel Process. Technol. **214**(106684), 29 (2021). https://doi.org/10.1016/j.fuproc.2020.106684

41. M. Aresta, in *The Carbon Dioxide Problem, in An Economy Based on Carbon Dioxide and Water, Potential of Large Scale Carbon Dioxide Utilization,* ed. by M.K. Aresta, Iftekhar, S. Kawi, (Springer, Switzerland AG, 2019)

42. Aresta, M. Carbon dioxide as chemical feedstock. Carbon dioxide as chemical feedstock. (Wiley-VCH, 2010), p. 394. https://doi.org/10.1002/9783527629916

43. M. Aresta et al., *An Economy Based on Carbon Dioxide and Water, Potential of Large Scale Carbon Dioxide Utilization*, ed. by M. Aresta, (Springer, Switzerland AG, 2019), p. 436. https://doi.org/10.1007/978-3-030-15868-2

44. S. Saeidi et al., Hydrogenation of CO2 to value-added products - a review and potential future developments. J. CO_2 Util. **5**, 66–81 (2014). https://doi.org/10.1016/j.jcou.2013.12.005

45. G.A. Olah et al., *Beyond Oil and Gas: The Methanol Economy*, 2nd edn. (Wiley-VCH, 2009), p. 334. https://doi.org/10.1002/9783527627806

46. A. Elmekawy et al., Technological advances in CO2 conversion electro-biorefinery: a step toward commercialization. Biores. Technol. **215**, 357–370 (2016). https://doi.org/10.1016/j.biortech.2016.03.023

47. M. Seemann, H. Thunman, Methane synthesis, in *Substitute Natural Gas from Waste*, ed. by M. Materazzi, P.U. Foscolo, (Academic Press, 2019), pp. 221–243. https://doi.org/10.1016/B978-0-12-815554-7.00009-X

48. C. Bassano et al., P2G movable modular plant operation on synthetic methane production from CO2 and hydrogen from renewables sources. Fuel **253**, 1071–1079 (2019). https://doi.org/10.1016/j.fuel.2019.05.074

49. K. Stangeland et al., CO2 methanation: the effect of catalysts and reaction conditions. Energy Procedia **105**, 2022–2027 (2017). https://doi.org/10.1016/j.egypro.2017.03.577

50. I. García-García et al., Power-to-gas: storing surplus electrical energy. Study of catalyst synthesis and operating conditions. Int. J. Hydrog. Energy **43**(37), 17737–17747 (2018). https://doi.org/10.1016/j.ijhydene.2018.06.192

51. J. Uebbing et al., Exergetic assessment of CO2 methanation processes for the chemical storage of renewable energies. Appl. Energy **233–234**, 271–282 (2019). https://doi.org/10.1016/j.apenergy.2018.10.014

52. J. Bremer et al., CO2 methanation: optimal start-up control of a fixed-bed reactor for power-to-gas applications. AIChE J. **63**(1), 23–31 (2017). https://doi.org/10.1002/aic.15496

53. S. Falcinelli, Fuel production from waste CO2 using renewable energies. Catal. Today **348**, 95–101 (2020). https://doi.org/10.1016/j.cattod.2019.08.041

54. J. Klankermayer, W. Leitner, Love at second sight for CO2 and H2 in organic synthesis. Science **350**(6261), 629–630 (2015). https://doi.org/10.1126/science.aac7997

55. F. Wang et al., Higher atmospheric CO2 levels favor C3 plants over C4 plants in utilizing ammonium as a nitrogen source. Front. Plant Sci. **11** (2020). https://doi.org/10.3389/fpls.2020.537443

56. R.C. Pullar et al., A review of solar thermochemical CO2 splitting using ceria-based ceramics with designed morphologies and microstructures. Front. Chem. **7**, 34 (2019). https://doi.org/10.3389/fchem.2019.00601

57. M. Levy et al., Solar energy storage via a closed-loop chemical heat pipe. Sol. Energy **50**(2), 179–189 (1993). https://doi.org/10.1016/0038-092X(93)90089-7

58. M. Aresta et al., The changing paradigm in CO2 utilization. J. CO_2 Util. **3–4**, 65–73 (2013). https://doi.org/10.1016/j.jcou.2013.08.001

59. S. Christy et al., Recent progress in the synthesis and applications of glycerol carbonate. Curr. Opin. Green Sustain. Chem. **14**, 99–107 (2018). https://doi.org/10.1016/j.cogsc.2018.09.003

60. M.M. Ramirez-Corredores et al., Radiation-induced chemistry of carbon dioxide: a pathway to close the carbon loop for a circular economy. Front. Energy Res. **8**(108), 17 (2020). https://doi.org/10.3389/fenrg.2020.00108

61. J.A. Rodríguez-Sarasty et al., Deep decarbonization in northeastern North America: The value of electricity market integration and hydropower. Energy Policy **152** (2021). https://doi.org/10.1016/j.enpol.2021.112210

62. National Research Council *Carbon Management: Implications for R&D in the Chemical Sciences and Technology*. (The National Academies Press, Washington, DC, 2001), p. 236. https://doi.org/10.17226/10153

63. A.P.M. Velenturf, P. Purnell, Principles for a sustainable circular economy. Sustain. Prod. Consum. **27**, 1437–1457 (2021). https://doi.org/10.1016/j.spc.2021.02.018
64. S. Fuss et al., Negative emissions—part 2: Costs, potentials and side effects. Environ. Res. Lett. **13**(6) (2018). https://doi.org/10.1088/1748-9326/aabf9f
65. J. Forster et al., Mapping feasibilities of greenhouse gas removal: Key issues, gaps and opening up assessments. Glob. Environ. Chang. **63** (2020). https://doi.org/10.1016/j.gloenvcha.2020. 102073
66. K. Dooley, S. Kartha, Land-based negative emissions: risks for climate mitigation and impacts on sustainable development. Int. Environ. Agreem: Polit. Law Econ. **18**(1), 79–98 (2018). https://doi.org/10.1007/s10784-017-9382-9
67. M. Brander et al., Carbon accounting for negative emissions technologies. Clim. Policy **21**(5), 699–717 (2021). https://doi.org/10.1080/14693062.2021.1878009
68. B. Metz et al., Carbon dioxide capture and storage. Special Report. Intergovernmental Panel on Climate Change (IPCC), (Cambridge University Press, UK 2005), p. 442
69. T.T.D. Cruz et al., Life cycle assessment of carbon capture and storage/utilization: from current state to future research directions and opportunities. Int. J. Greenh. Gas Control **108**(103309), 13 (2021). https://doi.org/10.1016/j.ijggc.2021.103309
70. F. Gassner, W. Leitner, Hydrogenation of carbon dioxide to formic acid using water-soluble rhodium catalyststs. J. Chem. Soc. Chem. Commun. **19**, 1465–1466 (1993). https://doi.org/10. 1039/C39930001465
71. Z.Z. Yang et al., CO2 capture and activation by superbase/polyethylene glycol and its subsequent conversion. Energy Environ. Sci. **4**(10), 3971–3975 (2011). https://doi.org/10.1039/c1e e02156g
72. S.M. Kim et al., Integrated CO2 capture and conversion as an efficient process for fuels from greenhouse gases. ACS Catal. **8**(4), 2815–2823 (2018). https://doi.org/10.1021/acscatal.7b0 3063
73. L. Liu et al., Integrated CO2 capture and photocatalytic conversion by a hybrid adsorbent/photocatalyst material. Appl. Catal. B **179**, 489–499 (2015). https://doi.org/10.1016/j. apcatb.2015.06.006
74. X. Wang, C. Song, Carbon capture from flue gas and the atmosphere: a perspective. Front. Energy Res. **8**, 24 (2020). https://doi.org/10.3389/fenrg.2020.560849
75. Z. Zhou et al., 2d-layered Ni–MgO–Al2O3 nanosheets for integrated capture and methanation of CO2. Chemsuschem **13**(2), 360–368 (2020). https://doi.org/10.1002/cssc.201902828
76. H.B. Vakil, P.G. Kosky, Design analyses of a methane-based chemical heat pipe, in *Proceedings of the 11th Intersoc Energy Conversion Engineering Conference* (New York, NY, September 12–17, 1976). AIChE. **1 SAE,** 659–664
77. A. Tripodi et al., Carbon dioxide methanation: design of a fully integrated plant. Energy Fuels **34**(6), 7242–7256 (2020). https://doi.org/10.1021/acs.energyfuels.0c00580
78. A. Álvarez et al., CO₂ activation over catalytic surfaces. ChemPhysChem **18**(22), 3135–3141 (2017). https://doi.org/10.1002/cphc.201700782
79. H. Yang et al., A review of the catalytic hydrogenation of carbon dioxide into value-added hydrocarbons. Catal. Sci. Technol. **7**(20), 4580–4598 (2017). https://doi.org/10.1039/c7cy01 403a
80. S. Verma et al., A gross-margin model for defining technoeconomic benchmarks in the electroreduction of CO2. Chemsuschem **9**(15), 1972–1979 (2016). https://doi.org/10.1002/cssc. 201600394
81. T. Burdyny, W.A. Smith, CO₂ reduction on gas-diffusion electrodes and why catalytic performance must be assessed at commercially-relevant conditions. Energy Environ. Sci. **12**(5), 1442–1453 (2019). https://doi.org/10.1039/C8EE03134G
82. J. Durrani, Can catalysis save us from our CO2 problem?. (2019). https://www.chemistryworld. com/news/can-catalysis-save-us-from-our-co2-problem/3010555.article
83. R.J. Lim et al., A review on the electrochemical reduction of CO2 in fuel cells, metal electrodes and molecular catalysts. Catal. Today **233**, 169–180 (2014). https://doi.org/10.1016/j.cattod. 2013.11.037

84. Y. Zhang et al., Mechanistic understanding of the electrocatalytic CO2 reduction reaction – new developments based on advanced instrumental techniques. Nano Today **31**, 100835 (2020). https://doi.org/10.1016/j.nantod.2019.100835

85. S. Baldwin et al., An assessment of energy technologies and research opportunities. (U.S. Department of Energy. Washington, DC. USA, 2015), p. 860

86. D. Sandalow et al., Carbon dioxide utilization roadmap 2.0. ICEF. November (2017), p. 30. https://www.icef.go.jp/platform/article_detail.php?article__id=171

87. S. Voitko et al., Decarbonisation of the economy through the introduction of innovative technologies into the energy sector, in *Proceedings of the International Conference on Sustainable, Circular Management and Environmental Engineering, ISCMEE 2021*. EDP Sciences. **255**(01016), p. 11. https://doi.org/10.1051/e3sconf/202125501016

88. W.M. Chen, H. Kim, Circular economy and energy transition: a nexus focusing on the non-energy use of fuels. Energy and Environment **30**(4), 586–600 (2019). https://doi.org/10.1177/0958305X19845759

89. S. Baldwin et al., Advancing clean electric power technologies, technology assessments, in *Quadrennial Technology Review—An Assessment of Energy Technologies and Research Opportunities* (U. S. Department of Energy, Washington, DC. USA, 2015), pp. 100–143

90. H. Ohno et al., Detailing the economy-wide carbon emission reduction potential of post-consumer recycling. Resour. Conserv. Recycl. **166** (2021). https://doi.org/10.1016/j.resconrec.2020.105263

91. L. Zhao et al., Drivers of household decarbonization: decoupling and decomposition analysis. J. Clean. Prod. **289** (2021). https://doi.org/10.1016/j.jclepro.2020.125154

92. M. Isik et al., Challenges in the CO2 emissions of the Turkish power sector: Evidence from a two-level decomposition approach. Util. Policy **70**(101227), 9 (2021) https://doi.org/10.1016/j.jup.2021.101227

93. Y. Ni et al., Novel integrated agricultural land management approach provides sustainable biomass feedstocks for bioplastics and supports the uk's 'net-zero' target. Environ. Res. Lett. **16**(1), 014023, 11 (2021). https://doi.org/10.1088/1748-9326/abcf79

94. A. Foss et al., NRIC integrated energy systems demonstration pre-conceptual designs. INL EXT-21–61413 Report. Idaho National Laboratory, National Reactor Innovation Center. Idaho Falls, ID. USA. April, 2021, p. 75. https://nric.inl.gov/wp-content/uploads/2021/06/NRIC-IES-Demonstration-Pre-conceptual-Designs-Report-1.pdf.

95. K. Derviş, S. Strauss, *The decarbonization paradox* (2021), p. 4. https://www.brookings.edu/opinions/the-decarbonization-paradox/. Accessed August 2021.

96. V. Smil, Energy myths and realities: bringing science to the energy policy debate. (AEI Press, Washington, D.C. USA, 2010), p 212

97. Deloitte, The 2030 decarbonization challenge: the path to the future of energy. Deloitte Global. (2020), p. 30. https://www2.deloitte.com/content/dam/Deloitte/global/Documents/Energy-and-Resources/gx-eri-decarbonization-report.pdf